James Carrol

METHODS OF STUDY IN QUANTITATIVE SOIL ECOLOGY:
POPULATION, PRODUCTION AND ENERGY FLOW

IBP HANDBOOK No. 18

Methods of Study in Quantitative Soil Ecology: population, production and energy flow

Edited by
J. PHILLIPSON

INTERNATIONAL BIOLOGICAL PROGRAMME
7 MARYLEBONE ROAD, LONDON NW1

BLACKWELL SCIENTIFIC PUBLICATIONS
OXFORD AND EDINBURGH

SBN 632 05680 0

First Published 1971

Distributed in the U.S.A. by
F. A. Davis Company, 1915 Arch Street
Philadelphia, Pennsylvania

Printed in Great Britain by
Burgess and Son (Abingdon) Ltd.
Abingdon, Berkshire
and bound by
The Kemp Hall Bindery, Oxford

Contents

Contents

Contributors

P. Berthet Laboratoire d'Ecologie Animale, Institut de Zoologie, Université de Louvain, Belgium

M.V. Brian The Nature Conservancy, Furzebrook Research Station, Near Wareham, Dorset, England

A.F.G. Dixon Department of Zoology, The University, Glasgow W.2, Scotland

C.A. Edwards Rothamsted Experimental Station, Harpenden, Herts., England

K.E. Fletcher Rothamsted Experimental Station, Harpenden, Herts., England

T.R.G. Gray Hartley Botanical Laboratories, University of Liverpool, Liverpool, England

G.C. Head Late of East Malling Research Station, East Malling, Maidstone, Kent, England
    Now at Technical Publications Unit, World Health Organization, Avenue Appia, 1211 Geneva, Switzerland

O.W. Heal The Nature Conservancy, Merlewood Research Station, Grange-over-Sands, Lancs., England

I.N. Healey Department of Zoology, King's College, University of London, Strand, London WC2, England

J. Holding School of Agriculture, University of Edinburgh, Edinburgh, Scotland

A. Macfadyen The New University of Ulster, Coleraine, County Londonderry, Northern Ireland

A. Medwecka-Kornas Nature Conservation Research Centre, Polish Academy of Sciences, Lubicz 46, Kraków, Poland

H.M. Nagel-de-Boois I.T.B.O.N., Arnhem, Netherlands

P.F. Newell Department of Zoology, Westfield College, University of London, London NW3, England

F.B. O'Connor Department of Zoology, University College London, Gower Street, London WC1, England

M. Oostenbrink Laboratory of Nematology of the Landbouwhogeschool, Binnenhaven, Wageningen, Netherlands

D. Parkinson Department of Biology and Environmental Sciences Centre, University of Calgary, Calgary, Canada

J. Phillipson Animal Ecology Research Group, Department of Zoology, University of Oxford, Oxford, England

J.E. Satchell The Nature Conservancy, Merlewood Research Station, Grange-over-Sands, Lancs., England

Foreword

The International Biological Programme, which started its operations in 1964 and lasts for a decade until 1974, is a world study of "biological productivity and human adaptability". It is divided into seven sections of which one of the largest, concerned with terrestrial productivity (PT), has sponsored this handbook. Most of the books in this series are concerned with methods for biological research but I would emphasize that their purpose is not to standardize or stabilize methodology, for most of the subjects concerned are growing points of biological science and the methodology is growing with the subject. It is most important that this process continues. Nevertheless, there are around the world large numbers of biologists who welcome advice about methods, particularly those methods which are calculated to give results from different climates and different ecosystems which will be inter-comparable one with another.

It is important, therefore, from time to time to bring together and to make available the up-to-date knowledge on methods and in the IBP this has generally been done through the aid of technical meetings composed of specialists in a particular sector of the programme. After discussion, the writing of a handbook is allocated to an appropriate group of authors under the guidance of a general editor. Such a process led to this volume. The technical meeting in this case was a large one and was organized jointly by UNESCO and IBP in November 1967 in Paris. Its scientific organizer, Dr John Phillipson, who is also editor of this handbook, is well known in the field of soil ecology and bio-energetics. He has recently moved from the Department of Zoology at the University of Durham in England to the University of Oxford where he has been heavily engaged over several years in developing, within the Department of Zoology, the Animal Ecology Research Group (the former Bureau of Animal Populations which was created and directed for many years by Charles Elton). Dr Phillipson's heavy commitments in this task have been responsible for a rather long delay between the design of the handbook at the technical meeting in 1967 and its publication now.

This is the fifth IBP handbook which provides guidance and advice on methods of research in section PT. The others are on primary production of forests (No. 2), on grasslands (No. 6), on the productivity of large herbivores (No. 7), and of other terrestrial animals (No. 13). It is hoped that all of these, and others which are currently in the press, will be widely used, not only by research workers within the programme, but by many other biologists. It is hoped also that the subject will progress so rapidly that his handbook will be out of date within a few years and that a revised and improved edition will then be called for.

E. B. WORTHINGTON

August 1971

Preface

The contributors to this volume came together at specially convened meetings immediately before and after the Unesco–IBP Symposium 'Methods of Study in Soil Ecology' which was held at Unesco Headquarters, Paris, in November 1967. It was at these meetings that the contents and order of presentation of the material proffered in this handbook were agreed.

The 'Proceedings' of the Paris Symposium were published by Unesco in 1970 under the title 'Methods of Study in Soil Ecology' and it might be felt by some that the 'Proceedings' should suffice as a guide to all those interested in pursuing the study of soil organisms within the framework of the International Biological Programme. However, the essence of most present-day international symposia is such that numerous, short, contributions are presented and participants frequently wish that some of the topics had been dealt with in greater depth. In consequence it was felt that an appropriate volume for the IBP Handbook series would be one entitled 'Methods of Study in Quantitative Soil Ecology: Population, Production and Energy Flow', hence this contribution. The topics covered are of a more restricted nature than those which appeared in the 'Proceedings', emphasis having been placed on the methodological aspects of studying production and energy flow through soil ecosystems, their component communities and species populations.

Communities of soil organisms, and their effective environment, form what is generally termed a 'soil ecosystem'; that is not to say that the system is 'isolated' or 'closed' for it is well known that there is considerable overlap of species inhabiting the soil proper and the litter and living vegetation above it. Nevertheless, in some instances it is convenient to treat a soil ecosystem as a unit and this treatment has been adopted in the present work. A soil ecosystem is extremely complex and it is not surprising that the contributors to this volume, each with expertise in his own field, approach soil ecosystems in different ways. Certain authors have adopted a holistic approach; the majority deal with specific groups of organisms, some of which may be regarded as permanent and others as periodic or temporary members of soil communities. No apologies are offered as each approach, and each group of organisms, requires special methods of study.

Inevitably some of the present chapters follow closely the articles published in 'Methods of Study in Soil Ecology', others are quite new. In a few instances there is slight overlap in chapter contents but no attempt was made to reduce this in that it was considered useful that users of the handbook could see for themselves how each contributor approaches problems in his specific field. The terminology of the different authors varies and reflects the large number of terms in common use, no attempt was made to standardize these although recommended terminology is given in IBP News 10.

This handbook attempts to provide, within a single volume, information about widely applicable methods of study in quantitative soil ecology, particular attention being given to populations, production and energy flow. No single approach or method is unanimously claimed superior to all others, information is provided but the final choice remains the prerogative of individual investigators.

JOHN PHILLIPSON

1

The Soil and its Total Metabolism

A. MACFADYEN

Introduction

The total amount of energy which is derived from sunlight and incorporated in plant tissues through photosynthesis varies remarkably little between terrestrial ecosystems in the same latitude. Further, since plant respiration as well as photosynthesis is increased at higher temperatures, the difference in the *net* primary production does not vary greatly with latitude.

A net dry matter production of 1 kg/m²/year (containing about 4,800 Kcal of energy) is a useful figure to remember and most normal terrestrial and fresh-water communities will neither fall short of this nor exceed it by a factor of more than two. (Some swamp systems may be up to eight times as productive however.) (Newbould, 1963), (Pearsall and Gorham, 1956).

This value of just under 5,000 Kcal is about 1·6% of the total photosynthetically useful solar energy at the earth's surface in temperate oceanic climates.

Although there is little variation in the value of net primary production, the partitioning of this energy between different parts of an ecosystem varies greatly, as does the calorific value of the living plant material which is supporting the photosynthetic activity (Table I).

It will be seen that in communities based on microscopic plants the ratio of stock to gross production is very small, and the proportion of net production consumed by herbivores is high. In woodland and forest the much larger stock of plant tissue respires more, so that proportionately less is available as secondary production. Of this again, a higher proportion is consumed by decomposer organisms. The two grassland examples show an intermediate value of stock, a low respiration value and a very high contribution to the decomposer part of the ecosystem.

It is only relatively recently that it has been realised what a high proportion of total organic matter in many terrestrial systems is present in the soil and litter and is decomposed there (Macfadyen, 1964). Clearly it is important to

1

TABLE I. Partition of primary production in different communities.
(Units are equivalents of kg dry matter/m²/year = 10 tonnes/Ha/year)

Community	Gross Primary production	Respiration	Net primary production			Mean Stock
			Total	Eaten	Decomposed	
Plankton: marine						
temperate	0·72	0·06	0·65	0·65	0·01	0·004
Algae: salt marsh	0·50	0·05	0·45	0·45	0·01	0·003
Spartina: salt marsh	1·17	0·10	1·07	0·07	1·00	1·06
Meadow: grazed	1·17	0·12	1·05	0·39	0·66	1·00
Beech wood: Denmark	2·35	1·00	1·35	0·95	0·40	15·5
Rain Forest: Ivory						
Coast	5·35	4·00	1·35	0·90	0·45	24·0

Derived percentages from above figures

Community	Gross primary production	Respiration	Net primary production			Mean stock S. prodn.
			Total	Eaten	Decomposed	
Plankton: marine						
temperate	100	9	90+	90	5	0·6
Algae: salt marsh	100	10	90+	90	5	0·6
Spartina: salt marsh	100	9	91	6	85	90
Meadow: grazed	100	10	90	33	56	85
Beech wood: Denmark	100	43	57	40	17	6,600
Rain Forest: Ivory						
Coast	100	75	26	17	9	4,570

Figures derived from Bray (1964), Macfadyen (1964, and in prep.), Maldague and Hilger (1963), Muller *et al.* (1960), and Odum and Smalley (1959).

study the process of decomposition of this matter both because much of the energy originating from primary production is released there, and also because it is only during this release that plant nutrients are made available for re-cycling to the system as a whole.

The metabolic activity of soil is the sum total of that of all the constituent soil inhabiting organisms. In the case of a few desert environments direct oxidation of soil organic matter occurs to a small extent. Also some decomposition is undoubtedly carried on by extracellular enzymes from micro-organisms, but this is presumably proportional to microbial population levels and is generally thought to be of minor importance (Hofmann, 1963).

Obviously, therefore, the greater part of total soil-metabolism could, in theory, be computed from the study of metabolism by all the organisms present. Unfortunately, although we now have reasonably accurate laboratory methods for measuring respiration of most organisms, we are a very long way from being able to relate these measurements to the environmental conditions actually obtaining in the field, to the physiological state of the organisms living there, and to the actual numbers of these organisms. In fact, the relatively fundamental question of estimating numbers of soil organisms is very far from being answered in many cases, whilst microbial metabolism in the field has not been measured directly at all.

Methods for the measurement of total soil metabolism

Most of the estimates of total soil metabolism made hitherto have been arrived at either by subtracting the energy consumed by herbivores from net primary production or by adding up estimates of the total metabolism of all the constituent organisms present. The first method is, theoretically, quite sound, but it compounds the errors from both these measurements and it is clearly desirable to attempt an independent measurement of the energy both going into the decomposer system and that leaving it. In a balanced system and in the absence of gains, losses or accumulation of organic matter in the soil these two should, of course, be equal.

The rate of addition of some components of organic matter to soil can be measured relatively easily. In woodlands, the leaf fall is measured directly; this has been done in many classical studies (e.g. Möller, Müller and Nielsen, 1954) and is a feature of current IBP programme (see Medwecka-Kornas, in prep.). However, appreciable amounts of organic matter are also added as faeces from herbivorous and carnivorous animals, and there are also gains and losses due to movements of animals between the soil and other parts of the system. In the case of grasses and many other plants the accurate measurement of litter production is far from easy, whilst the measurement of the contribution made by roots to soil organic matter is extremely difficult, and has in fact hardly been seriously attempted yet.

In order to relate litter fall to corresponding expected carbon dioxide production a conversion factor must either be determined or assumed. Theoretically, in the case of all carbon compounds, $E_c = \dfrac{H}{C}$ where:—

E_c = energy content per g Carbon
H = heat of combustion per g substance in kcal
C = carbon content of substance (expressed as a decimal)

Since each 22·4 litres of CO_2 are equivalent to 12 g of Carbon, if E_{co_2} = energy liberated in the evolution of 1 litre of $CO_2 = \dfrac{E_c \times 12}{22·4} \times 0·535\ E_c$

$$\text{then } E_{co_2} = \frac{0·535\ H}{C}$$

For example, in the case of pinewood we have $H = 4·785$ kcal/g, $C = 0·482$ (48·2%) $\therefore E_c = 9·92$ kcal and $E_{co_2} = 5·3$ kcal/G; corresponding figures for starch or cellulose would be $H = 3·74$, $C = 0·40$, $E_c = 9·36$ and $E_{co_2} = 5·0$.

Quite different methods have also been used to obtain comparative data on litter decomposition and to assess the relative contributions made to the decomposition process by different groups of organisms. In particular bags of plastic fibre such as 'Nylon' and 'Terylene' containing weighed amounts of litter have been exposed in the field for controlled periods and the losses in weight or in surface area of punched litter discs have been measured and related to the litter type, mesh size of the bag and to the vegetation and environmental factors in the exposure site (Bocock and Gilbert, 1957; Bocock *et al.*, 1960; Crossley and Hoglund, 1962; Edwards and Heath, 1963; Witkamp and Olson, 1963). Although results of such methods have been shown to correlate with microbial counts and litter carbon dioxide output (Witkamp, 1966a) they cannot, of course, be used directly to give data for comparison with primary and secondary production of litter fall.

The second possibility is to attempt to add up the metabolism under field conditions of all the organisms present. Work on these lines is progressing all the time but the more we know, the greater seems to be the complexity of the task. For instance it has been shown that metabolic activity of soil arthropods depends on age, sex, reproductive condition and season (Phillipson, 1960a, 1960b, 1967; Phillipson and Watson, 1965). There is some disagreement between workers in this field as to the effect of temperature on metabolism. Some have measured Q_{10} factors as high as five or six (Berthet,

1964), others around 2 (Krogh, 1914; Nielsen and Evans, 1960; Jørgensen, 1916) and (Webb, in prep.) whilst other authors maintain that the temperature effect is insignificant (Phillipson, 1967).

The effects of other environmental variables including moisture and soil gas concentrations have hardly been measured as yet.

Further, there are whole groups of important soil organisms for which it is simply not yet practicable to assess metabolic activity, and this applies especially to the most important group of all, the micro-organisms.

Clearly, if we are not to wait for a generation or more whilst these difficulties are resolved, those of us who are interested in (a) assessing the role of soil in metabolic studies of whole ecosystems and (b) attempting to gain even a first idea of the relative importance of particular groups of soil-organisms as compared with the soil biota as a whole, must try to measure total soil metabolism.

It follows, therefore, that as at least a partial check on the technique of summing individual organisms, metabolism and the 'litter fall' method of measuring the input to the soil, it is desirable to try to measure the overall decomposition rate of organic material. The only way in which this can be done is by measuring the entire metabolic activity of the soil. In theory, of course, metabolic activity can be measured directly by calorimetry or indirectly through oxygen consumption or carbon dioxide output. At least two of these quantities are needed in order to estimate the respiratory quotient, which is likely to be abnormal especially under anaerobic conditions. In practice soil calorimetry has hardly been attempted yet (but see Lemée *et al.*, 1958).

There has been a long history of soil respirometry based on gas analysis (Macfadyen, in prep.); in most cases carbon dioxide output rather than oxygen uptake has been measured. Earlier authors (Russell and Appleyard, 1925; Waksman and Starkey, 1924) were more interested in the composition of soil atmosphere than the production of gas, and tended to use soil which had been transferred to the laboratory. They ignored the effects of disturbance and of factors such as temperature, humidity and carbon dioxide levels on soil respiration; all of these have now been shown to have considerable effects. Many workers have removed soil samples to the laboratory and, after various preparatory treatments, measured respiration in conventional respirometers (Waksman and Starkey, 1924; Gaarder, 1957) or in electrolytic respirometers (e.g. Swaby and Passey, 1953; Birch and Friend, 1956).

On the other hand authors have made field measurements by covering soil *in situ* with a bell jar or other open ended container; carbon dioxide output is, then, measured in one of the following ways:

(a) Estimation of increase in carbon dioxide content of the enclosed air by periodical analysis of small samples extracted at intervals (e.g. Köpf, 1952).

(b) Continuous absorption of carbon dioxide in alkali and determination of the amount absorbed at the end of the experiment either by titration (Romell, 1927; Lundegardh, 1927; Lieth and Ouellette, 1962; Walter and Haber, 1957; Schultze, 1967) or gravimetrically (Monteith *et al.*, 1964; Howard, 1966).

(c) Continuous circulation of air from the enclosure by means of a pump together with absorption of carbon dioxide from the gas stream in alkali. This is followed by titration or gravimetric estimation (Wallis and Wilde, 1957).

A comparison between results obtained from litter fall and from soil respiration is given in Table II. It is at once obvious that soil respiration methods produce higher estimates than do direct litter fall measurements and that only the figures due to Witkamp (1966b) and to Macfadyen (in prep.) appear at all reasonable. One reason for this is that soil respiration figures include carbon dioxide derived from roots which may amount to 50% of the total in woodland soils (Bray and Gorham, 1964; Macfadyen, in prep.). When allowance is made for root respiration, soil respiration and litter fall, figures approximate much more closely but the former are probably still high. This is probably due to simulation of soil metabolism by the operator, a factor which has been largely eliminated by the latest improvements in technique (Macfadyen, in prep.; Brown and Macfadyen, 1969).

It is hardly surprising that mechanical disturbance should increase respiration in view of the well known effects of almost any kind of disturbance in promoting microbial activity (e.g. Dobbs and Hinson, 1960). It is even possible that an increased air-flow may stimulate respiratory activity to an even greater extent because the values quoted by Wallis and Wilde are almost certainly an order of magnitude too high. This may be related to the effect of carbon dioxide on microbial activity (Macfadyen, in press).

A major improvement in such methods is due to Witkamp (1963, 1966a, 1966b) who probably first suggested using open-ended cylinder ('inverted box method') which could be left in the soil for long periods (weeks) in order that the effects of disturbance might be reduced. These cylinders are then capped by an air-tight cover for a short while (an hour or two) during which carbon dioxide derived from the soil is absorbed in alkali and its quantity is estimated by titration.

TABLE II. Comparisons of energy and dry matter equivalents of litter as estimated by direct measurement of litter fall compared with estimates from soil respiration.
(All figures represent total quantities per square metre per annum. All conversions are based on an equivalent of 1 kg dry matter$=4,800$ kcal$=700$ litres of CO_2).

Type of estimate		Author	Dry matter kg	Energy kcal	Carbon dioxide litres
(a) TEMPERATE OAKWOODS					
Litter fall	mor	Drift (1963)	354	1,410	248
,,	mull	,,	332	1,320	232
,,	mull	Bray and Gorham (1964)	440	1,750	308
,,	mull	Witkamp (1966b)	538	2,150	377
Soil respiration		,, ,,	1,110	3,030	775
,,		Macfadyen (in prep.)	500	2,000	350
,,		*Feher (1933)	4,430	17,700	3,100
,,		*Lundegardh (1927)	11,100	44,200	7,760
,,		†*Witkamp and Drift (1961)	2,310	9,250	1,620
,,		*Wallis and Wilde (1957)	35,400	142,000	24,800
(b) TROPICAL FOREST					
Litter fall		Nye (1961)	1,055	4,220	740
Soil respiration		Maldague and Hilger (1963)	3,190	13,700	2,230

(c) MISCELLANEOUS TEMPERATE HABITATS:
SOIL RESPIRATION
(All corrected from summer readings)

Habitat	Country	Author	Method	Dry matter kg	Energy kcal	Carbon dioxide litres
*Grassland	Norway	Gaarder (1957)	Warburg 25°C	1,330	5,660	920
*Arable	England	Russell (1950)	,,	673	2,890	471
* ,,	Germany	Köpf (1952)	Gas samples	955	4,300	700
*Beech	Denmark	Romell (1927)	Warburg	2,130–3,180	9,150–13,620	1,490–2,220

* These authors quote summer-time results only. Their results are multiplied by 0·5 to obtain very approximate mean annual figures.
† Revised figures supplied by Dr Minderman from I.T.B.O.N. records.

Results

The main factors affecting the amounts of carbon dioxide absorbed in such an apparatus are likely to be:—

(1) *Respiration from plant roots.* This factor is a major unknown in such studies, and it urgently requires investigation. Although root respiration measurements from laboratory plants under comparable physical conditions would be of some value, these are hardly realistic because roots in nature are inseparable from their rhizosphere organisms and their activity must frequently be influenced by these, and by the physical factors in the field. This is clearly a type of study in which the use of radio-active isotopes is desirable, and some work of this kind has already been started. In general botanical ecologists and physiologists have done little work on roots and few reliable estimates of their importance are available.

Bray (1963) (see also Bray and Gorham, 1964) estimated that woodland root production is usually about 20% of above ground tree production (i.e. 17% of the total) whilst Minderman (1967) obtained an almost identical figure (16·4%) for *Pinus niger* var *austriaca* growing in sandy conditions. Bray, Lawrence and Pearson (1959) calculate an equivalent ratio of 30% for maize. Chew and Chew (1965) estimated a ratio of about 12% for the creosote bush in a desert scrub community, whilst Bliss's (1966) figures suggest a root production in Alpine sedge meadow approaching 50%. It seems likely that most ratios lie between 15% and 50% with woodlands towards the lower end and grassland towards the upper end. Even if the production of tree roots should lie at the lower end of this range, however, it is likely that the relatively lower activity of their roots is more than counterbalanced by the greater biomass and respiration of trees which, on an area basis, greatly exceeds those of grasses. Referring back to Table I, it will be seen that the Danish beechwood respires at a rate equivalent to about 1 kg of dry matter/m²/year, whilst the figure for decomposition is 0·4 kg; a ratio of 2·5: 1. In the grassland the equivalent ratio at 1: 5·5 is reversed.

The effect of this very different distribution of gross primary production on the expected sources of carbon dioxide emerging from the soil surface can be seen in Table III.

Although the precise figures in this table are very approximate, in the present context of assessing the utility of soil carbon dioxide measurements as a measure of decomposition activity, it is clear that figures obtained will be much more trustworthy in the case of grassland (and heath) areas than in

TABLE III. Very approximate estimation of relative importance of root respiration in total below-ground metabolism. (Units are thousands of $kcal/m^2/annum$).

	Beech	Grass
Total plant respiration (Table I)	4·0	0·47
Per cent of above due to roots	17%	30%
Respiration due to roots	0·68	0·14
Decomposition (Table I)	1·60	2·62
Decomposition plus roots = total sub-surface metabolism	2·28	2·76
Decomposition as percentage of total sub-surface metabolism	70%	94·9%

woodlands. In the latter case up to half the carbon dioxide might originate from tree roots and the soil carbon dioxide figures would require a large subtractive correction. In other habitats this correction might be almost negligible but evidence has been obtained (Macfadyen, in prep.) that root respiration in grass and *Calluna* heath becomes proportionately more important in summer than in winter.

(2) *Transient effects* due to changes in atmospheric pressure etc. Although serious errors might be anticipated from this source, they can, in fact, be allowed for.

(3) *The metabolic activity of the soil organisms*, as modified by temperature and other physical factors. Clearly it is important to determine the effect of such modifying factors on metabolism and to correct for the conditions obtaining at the time when measurements are made. The most important of these factors are:

Temperature. Under natural conditions this follows the relation: $Q_{10} = 2$.

Humidity. Reduction of metabolic activity to about 50% under very dry conditions in sandy *Calluna* heath is possible. Over most of the year this factor will not be important in Western Europe but should be investigated in other areas with more extreme climates.

Carbon dioxide. Although possibly not of general importance, a significant effect at CO_2 concentrations which do occur naturally has been found in the *Calluna* habitat mentioned above (Macfadyen, in press; Burges and Fenton, 1963).

Conclusions

Until recently the decomposer organisms in soil were largely ignored by ecologists. Most recent work demonstrates clearly the relatively great importance of these forms as regards (1) the range of organisms present, (2) the variety of biochemical changes which take place, and (3) the relative proportion of secondary production and heterotrophic activity occurring there. It is universally agreed by soil microbiologists and even by soil zoologists that the complete and simultaneous compilation of metabolic budgets for all organisms in a given soil type over a meaningful area and time period is an exercise which will not be completed in the foreseeable future. All we can hope to do at present is to work with representative organisms over realistic ranges of conditions and try to assess their relative importance in comparison with the whole soil ecosystem. Further, we need a means to compare the decomposer systems as a whole, both with those of other soils and also with other components of the ecosystem. For the first of these purposes a comparative measure of soil metabolism is probably adequate; quite realistic comparisons have even been made using dried and re-wetted soil samples in Warburg respirometers at standard temperatures (e.g. Chase and Gray, 1953).

However, if we are to fit the activities of soil decomposers into ecosystems as a whole, we must either measure their true activity in the field or we must determine corrections from which such data can be derived.

It appears that only one method has so far been developed which shows promise of meeting the more stringent criteria. Fortunately the method, described here, is simple and can be readily replicated. It demands careful observance of some elementary precautions in its use and results still require correction for the effects of plant root respiration. These are relatively insignificant in some types of habitat, but until methods for measuring this factor have been developed, figures for soil carbon dioxide output will be subject to a subtractive correction of up to 50% from this source in some habitats.

Acknowledgements

I am greatly indebted to contributors at the Paris symposium who made constructive criticism of the above paper. In particular Professor Ghilarov, Dr. Loissant, Professor Newbould and especially Dr. Minderman provided important information, much of which has been incorporated in this revised text.

References

BERTHET P. (1964). *L'activité des oribatides d'une Chenaie*, 152 pp. (Mem. Inst. Roy. sci. nat. Belg.).

BIRCH L.C. & FRIEND M.T. (1956). Humus decomposition in East African soils. *Nature Lond.* **178**, 500–501.

BLISS L.C. (1966). Plant productivity in Alpine microenvironments. *Ecol. monogr.* **36,** 125–155.

BOCOCK K.L. & GILBERT O.J. (1957). The disappearance of leaf litter under different woodland conditions. *Plant and Soil,* **9,** 179–185.

BOCOCK K.L., GILBERT O.J., CAPSTICK C.K., TWINN D.W., WAID J.S. & WOODMAN M.J. (1960). Changes in leaf litter when placed on the surface of soils with contrasting humus types. I. Losses in dry weight of oak and ash litter. *J. Soil Sci.* **11,** 1–9.

BRAY J.R. (1962). The primary productivity of vegetation in central Minnesota, U.S.A., and its relationship to chlorophyll content and albedo. In: *Die Stoffproduktion der Pflanzendecke.* Fischer, Stuttgart. 102–116.

BRAY J.R. (1963). Root production and the estimation of net productivity. *Canad. J. Bot.* **41,** 65–72.

BRAY J.R. (1964). Primary consumption in three forest canopies. *Ecology,* **45,** 165–167.

BRAY J.R. & GORHAM E. (1964). Litter production in forests of the world. In: *Adv. ecol. Res.* J.B. Cragg (ed.). **2,** 101–157.

BRAY J.R., LAWRENCE D.B. & PAARSON L.C. (1959). Primary production in some Minnesota terrestrial communities for 1957. *Oikos,* **10,** 38–49.

BROWN A. & MACFADYEN A. (1969). Relation between soil carbon dioxide output and small scale vegetation pattern in a *Calluna* heath. *Oikos.*

BURGES A. & FENTON E. (1953). The effect of carbon dioxide on the growth of certain soil fungi. *Trans. Brit. mycol. Soc.* **36,** 104–108.

CHASE F.E. & GRAY P.H.H. (1953). Use of the Warburg respirometer to study microbial activity in soils. *Nature, Lond.* **171,** 481.

CHEW R.M. & CHEW A.E. (1965). The primary productivity of a desert-shrub (*Larrea tridentata*) community. *Ecol. Monogr.* **35,** 355–375.

CROSSLEY D.A. JR. & HOGLUND M.P. (1962). A litter bag method for the study of micro-arthropods inhaviting leaf litter. *Ecology,* **43,** 571–574.

DOBBS C.G. & HINSON W.H. (1960). Some observations on fungal spores in soil. In: *The Ecology of Soil Fungi.* D. Parkinson & J.S. Waid (eds.). Liverpool Univ. Press, 33–42.

DRIFT J. VAN DER (1963). The disappearance of litter in mull and mor in connection with weather conditions and the activity of the macrofauna. In: *Soil Organisms.* J. Doeksen & J. van der Drift (1963). North Holland Publishing Company, Amsterdam. 125–133.

EDWARDS C.A. & HEATH G.W. (1963). The role of soil animals in breakdown of leaf material. In: *Soil Organisms.* J. Doeksen & J. van der Drift (1963). North Holland Publishing Company, Amsterdam. 76–84.

FEHÉR D. (1933). *Mikrobiologie des Waldbodens.* Berlin, 272 pp.

GAARDER T. (1957). Studies in soil respiration in western Norway, the Bergen district. *Univ. i. Bergen. Arbok.* **3,** 24.

GIMMINGHAM (1964). Dwarf shrub heaths. In: *Vegetation of Scotland.* Burnet (ed.) Oliver & Boyd. 232–288.

HOFMANN F. (1963). The importance of the measurement of enzyme activity in botanical and agricultural chemistry. In: *Methods of Enzymatic Analysis.* H.U. Bergmeyer (ed.). Academic Press. 720–723.

HOWARD P.J.A. (1966). A method for the estimation of carbon dioxide evolved from the surface of soil in the field. *Oikos,* **17,** 267–275.

JØRGENSEN N.R. (1916). *Undersøgelser over Frequensflader og Korrelation.* Copenhagen. (Quoted in Nielsen E.T. & Evans D.G. (1960).)

KÖPF H. (1952). Laufende messung der Bodenatmung im Freiland. *Landwirtschaftliche Forschung,* 4, 186–194.

KROGH A. (1914) The quantitive relation between temperature and standard metabolism in animals. *Int. Z. phys.-chem. Biol.* 1, 491–508.

LEMÉE G., LOISSANT, P., METTAUER, H. & WEISSBECKERR, R. (1958). Récherches préliminaires sur les charactères bio-chimiques de l'humus dans quelques groupements forestiers de la Plaine d'Alsace. Angew. *Pflanzensoziologie,* 15, 93–101.

LIETH H. & OUELLETTE R. (1962). Studies on the vegetation of the Gaspé peninsula. II. The soil respiration of some plant communities. *Canad. J. Bot.* 40, 127–140.,

LUNDEGARDH H. (1927). Carbon dioxide evolution of soil and crop growth. *Soil Sci.* 23, 417–450.

MACFADYEN A. (1963). *Animal Ecology: Aims and Methods.* 2nd edn. Pitman, London. 344 pp.

MACFADYEN A. (1964). Energy flow in ecosystems and its exploitation by grazing. In: *Symposium on Grazing.* D.J. Crisp (ed.). British Ecological Society. Blackwell, Oxford.

MACFADYEN A. (in prep.). Soil carbon dioxide production as an index of soil metabolism.

MACFADYEN A. (1970). Simple techniques for measuring and maintaining the proportion of carbon dioxide in air for use in ecological studies of soil respiration. *Soil Biol. Biochem.* 2, 9–18.

MACFADYEN A. (in press). The inhibitory effect of carbon dioxide on the metabolism of certain soils. *Soil Biol. Biochem.* 3.

MALDAGUE M.E. & HILGER F. (1963). Observations faunistiques et microbiologiques dans quelques biotopes forestiers equitoriaux. In: *Soil Organisms.* J. Doeksen & J. van der Drift (eds.). North Holland Publishing Company.

MEDWECKA-KORNAS A. (in prep.). Litter production. *UNESCO Symposium on Methods of Study in Soil Ecology.*

MÖLLER C.M., MÜLLER D. & JÖRGEN NIELSEN (1954). The dry matter production of European Beech. *Det Forstlige Forsøgsvaesen i Danmark,* 21, 253–335.

MONTEITH J.L., SZEICZ G. & YABUKI K. (1964). Crop photosynthesis and the flux of carbon dioxide below the canopy. *J. appl. Ecol.* 1, 321–337.

MÜLLER D. & JÖRGEN NIELSEN (1965). Production brute, pertes par réspiration et production nette dans la forêt ombrophile tropicale. *Det Forstlige Forsøgsvaesen i Danmark,* 29, 69–160.

NEWBOULD P.J. (1963). Production Ecology. *Science Progress,* 51, 91–104.

NIELSEN E., TETENS & EVANS D.G. (1960). Duration of the pupal stage of *Aedes taeniorhynchus,* with a discussion of the velocity of development as a function of temperature. *Oikos,* 11, 200–222.

NYE P.H. (1961). Organic matter and nutrient cycles under moist tropical forest. *Plant and Soil,* 13, 333–346.

ODUM E.P. & SMALLEY A.E. (1959). Comparison of population energy flow of a herbivorous and a deposit-feeding invertebrate in a salt marsh ecosystem. *Proc. nat. Acad. Sci. Wash.* 45, 617–622.

PEARSALL W.H. (1945). Leaf fall in Hertfordshire woodlands. *Trans. Herts. Nat. Hist. Soc.* **22**, 97–98.

PEARSALL W.H. & GORHAM E. (1956). Production Ecology. 1. Standing crops of natural vegetation. *Oikos*, **7**, 193–201.

PHILLIPSON J. (1960a). A contribution to the feeding biology of *Mitopus morio* (F) (Phalangida). *J. Anim. Ecol.* **29**, 35–43.

PHILLIPSON J. (1960b). The food consumption of different instars of *Mitopus morio* (F). (Phalangida) under natural conditions. *J. Anim. Ecol.* **29**, 299–307.

PHILLIPSON J. (1967). Studies on the bioenergetics of woodland *Diplopoda. Ekologia Polska* (in press).

PHILLIPSON J. & WATSON J. (1965). Respiratory metabolism of the terrestrial isopod *Oniscus asellus* L. *Oikos*, **16**, 78–87.

ROMELL L.G. (1927). Studier över kolsyrehushalt ovrigen i moss-rik tallsog. *Medd. Stats. Skogsförsöksanstalt*, **24**, 1–56.

RUSSELL E.J. (1950). *Soil Conditions and Plant Growth*, 8th edn. recast and rewritten by E.W. Russell (London, Longmans). 636 pp.

RUSSELL E.J. & APPLEYARD A. (1925). The atmosphere of the soil: its composition and the causes of variation. *J. agric. Sci.* **7**, 1–48.

SCHULZE E.D. (1967). Soil respiration of tropical vegetation types. *Ecology*, **48**, 642–653.

SWABY R.J. & PASSEY I.B. (1953). A simple macrorespirometer for studies in soil microbiology. *Austr. J. agric. Res.* **4**, 334.

WAKSMAN S. & STARKEY R.L. (1924). Microbiological analysis of soil as an index of soil fertility. 7. Carbon dioxide evolution. *Soil Science.* **17**, 141.

WALLIS G.W. & WILDE S.A. (1957). A rapid method for the determination of carbon dioxide evolved from forest soils. *Ecology*, **38**, 359–361.

WALTER H. & HABER W. (1957). Über die Intensität der Bodenatmung mit Bemerkungen zu den Lundergardschen Werten. *Ber. deut. botan. Ges.* **70**, 257–282.

WEBB N. (in prep.). *Studies on the Ecology of Soil Oribatei.* Ph.D. Thesis. University College of Swansea.

WITKAMP M. (1963). Microbial populations of leaf litter in relation to environmental conditions and decomposition. *Ecology*, **44**, 370–377.

WITKAMP M. (1966a). Decomposition of leaf litter in relation to environmental conditions, microflora and microbial respiration. *Ecology*, **47**, 194–201.

WITKAMP M. (1966b). Rates of carbon dioxide evolution from the forest floor. *Ecology*, **47**, 492–494.

WITKAMP M. & DRIFT J. VAN DER (1961). Breakdown of forest litter in relation to environmental factors. *Plant and Soil*, **15**, 295–311.

WITKAMP M. & OLSON J.S. (1963). Breakdown of confined and non-confined oak litter. Radiation Ecology Section, Oak Ridge. Private circular, 18 pp.

2

Plant Roots

G.C. HEAD

Introduction

The net primary production of an ecological community is the total quantity of organic matter synthesized by plants as a result of photosynthesis and mineral absorption during a specified time, usually a year. Production may be measured in a number of ways (Odum and Odum, 1959) including the measurement of biomass change, oxygen evolution, carbon dioxide utilization or chlorophyll content of the community, but of these only the change of biomass can have direct application to root systems.

It has been stressed repeatedly in textbooks of ecology that biomass and production should not be confused. It is the *change* in biomass that indicates production and it must be emphasized that measurement of the mass of the standing vegetation is not sufficient. Many of the plant parts that grow during the year are not permanent and the standing vegetation measurement must be corrected for the dry weight of those parts that become detached, leached out, excreted, eaten or harvested from the stand during the course of the production period.

The aerial parts of plants are relatively easy to measure and, knowing which parts are likely to become detached, appropriate steps may be taken to collect samples of the budscales, flowers, pollen, fruit, leaves, broken branches or whatever the parts may be. Knowledge of root life-history is very limited by comparison, except perhaps in the cases of a few of the more important crop plants, and even in these cases very little is known about the parts that become detached or are eaten away during the course of a year.

Biomass estimation and root life-history

It is not possible in this publication to give precise information on sampling procedures, as these will vary with the type of vegetation to be studied. The number and size of samples taken at any one time can be decided only after an initial survey of root distribution and density. The whole soil volume occupied by roots should be sampled and adequate samples of all types of root should be obtained. With woody plants, for example, growth occurs by the thickening of existing roots as well as by the addition of new roots and both types should be sampled.

Annual biomass samples are unlikely to be sufficient for root production studies and the decision on the frequency of sampling will depend to a large extent on the length of life of individual roots and the periodicity of new root growth.

As this handbook is primarily concerned with the amount of energy available to the multitudes of soil organisms which feed directly or indirectly on root material, the breakdown of figures for root production and biomass into broad categories, e.g. woody roots, new white extension roots, small rootlets, mucilage or excreted materials, and dead tissues will be useful for the soil biologist. Figures for total root system biomass of woody plants (and perhaps root production) are likely to be heavily weighted by the mass of the woody roots, which, although they form part of the energy store, are not generally the most readily available food materials.

It is obvious from the above remarks that any attempt to estimate root production must be based on whatever knowledge of roots is available, and where such knowledge is lacking no progress will be possible until the deficiency has been corrected.

More information on root life-history and periodicity of growth will, at least, give an idea of what is being missed in biomass samples and should enable the sampling programme to be rationalized and should help considerably in the interpretation of the samples obtained.

At East Malling Research Station the roots of several species of deciduous fruit plants have been studied for many years and it will be useful at this point to recount briefly some of the results in order to draw attention to points that should be borne in mind in future studies and to indicate some of the differences that occur between species.

The new white extension roots of fruit plants vary in diameter up to 2–3 mm and also vary in growth rate up to 7–8 cm per week. Roots of cherry trees, one of the more vigorous species, extend up to 3 m from the base of the trunk in the first growing season after planting. The cortex of the white roots of apple, cherry, plum and quince roots begins to turn brown after 1–3 weeks in the summer, as the endodermis and exodermis become suberized, but may remain white for several months during the winter (Head, 1966). The extension roots develop lateral branches after 2–3 weeks and many of these do not grow longer than a few centimetres. Subsequently the lateral roots may branch many times to form the typical networks of short absorbing rootlets. These short rootlets which are white or sometimes almost clear when young, Fig. 1a, often become very dark after a few weeks, but do not lose their cortex as described for larger roots, Fig. 1b. Individual small rootlets may rot away after several months or sometimes not until 2–3 years have passed, or the whole network may decay at one time, as in Fig. 1c. In contrast, many of the lateral roots on black currant and strawberry remain white and apparently unchanged for twelve months or more.

Browning of the cortex of apple tree roots is quickly followed by decay and the dispersal of the decomposing tissue is largely a result of feeding by species of the soil fauna. The central vascular cylinder, which remains after the cortex has rotted away, may be only about half the diameter of the original root and Rogers (1969) has shown, with extension roots of apple, that the cortex which rots away represents half the dry weight of the root at that stage. Many such conducting roots remain alive for several years with no increase in diameter. Fig. 2a shows a new white extending root beside a woody root and Fig. 2b, the same root $2\frac{1}{2}$ years later with just the narrow vascular cylinder remaining.

Once the cortex has rotted away, some of the extension roots, and some of the laterals commence secondary thickening, sometimes before they are 12 months old and become permanent woody roots. Thickening in subsequent years is often irregular, some roots thickening for a number of consecutive years and then showing no further radial increase for several more years (Head, 1968). Fig. 2 shows, in addition to the new white root, a woody root which increased in diameter in 1963 and 1964 (the years before Fig. 2a) but showed no further increase in 1965 and 1966 (the years between Fig. 2a and Fig. 2b). Thus it is clear that annual rings in woody roots should not be used as a guide to root age.

Methods used for estimating root biomass and production

All estimates of root production in previous studies have been based on the change in root biomass. Estimates of this quantity may be obtained by complete excavation of the root system (Rogers and Vyvyan, 1934), by sampling into trench sides either with a needle-board (Coker, 1958) or by washing-out roots from sample monoliths (Greenland and Koval, 1960). Alternatively soil and roots may be sampled by means of auger borings and various types of mechanical devices have been designed to ease the labour of auger sampling (Jutras and Tarjan, 1965). Each of these methods has been considered and reviewed by Kolesnikov (1962) and by Schuurman and Goedewaagen (1965).

The possibility that root biomass can be estimated from a knowledge of the biomass of part or all of the aerial parts has been investigated by Baskerville (1965) who found that roots formed 21·3–23·3% of the total dry weight of immature 42-year-old balsam fir trees ranging from one to 10 inches diameter at breast height, the proportion of root decreasing regularly with tree size. The decrease in root : shoot ratio with increasing tree size was confirmed with loblolly pine by Monk (1966). Rogers and Vyvyan (1934) showed with apple trees that the root : shoot ratio was fairly constant on a given soil type irrespective of tree size but varied considerably on different soils. Maggs (1961) examined the distribution of new growth between the various organs of young apple trees and found that differences in variety, water supply, mineral nutrition and illumination altered the over-all distribution of increment to leaves, stem and root. Bray (1963) reviewed data available from the literature and found considerable variation in the ratio of root production to shoot production, the ratio changing from 0·15 to 5·50 for herbaceous plants and from 0·10 to 0·32 for arboreal species. Clearly the root production of mixed communities or different species cannot readily be compared by calculations based on measurements of the tops alone.

Much of the work on production in natural communities has been carried out in the U.S.S.R. and many of the results from the Soviet investigations and from other countries have been reviewed by Rodin and Bazilevich (1965). The methods suggested for further studies of root production by Remezov, Rodin and Bazilevich (1963) are all based on estimation of changes in root biomass with corrections for natural loss of roots during the year.

For plants with woody roots Remezov *et al.* recommend that the total root biomass should be estimated by excavating the roots of specimen plants in the community and then multiplying this quantity by the number of specimens per unit area. Losses of root material from woody root systems are very difficult to assess and the only attempt to estimate these losses has been by using the laborious methods of Orlov (1955) who examined monoliths of soil with needle and forceps under a magnifying glass, picking out by hand living roots, dead roots and also recognizable fragments of decaying root material. Orlov's investigations in 25-year-old and 50-year-old *Picea excelsa* forests showed that the soil down to 200 cm contained 4·0 and 1·2 metric tons per hectare dry weight of living rootlets <0·3 mm in diameter, respectively, and that 50% and 20% respectively of the mass of small rootlets died off each year. Remezov (1959) used similar methods in a 50-year-old oak forest and found a decrease in dry weight of rootlets from 0·8 tons per hectare in July to 0·5 tons per hectare in August and considered that this confirmed Orlov's conclusion that up to 50% of the small rootlets may die each year.

For the herbaceous members of plant communities Remezov, Rodin and Bazilevich (1963) state that annual root loss has not been studied and suggest that the annual root production of perennial species should be taken as one-third of the biomass of underground organs. In communities where root biomass remains fairly constant it is clear that annual production more or less equals annual loss. Dahlman and Kucera (1965) took root biomass samples four times a year in a Central Missouri prairie and the difference between the highest and lowest values indicated that 25% of the root system as a whole would be replaced each year. Although turnover values were similar in the different horizons studied, down to 34 in, it was recognized that certain parts of the root system might be replaced more or less rapidly. These estimates of root loss are in broad agreement with the results of Weaver and Zink (1946) that roots of prairie grasses live on average for 2 or 3 years, although Weaver and others (Troughton, 1957) have shown that length of life of roots is much affected by grazing and other management treatments.

Most natural communities contain a mixture of woody plants, herbaceous perennials and annual plants, and in these cases it is necessary in production studies to determine the proportion of the root biomass belonging to plants of each type. Tracer techniques are valuable in labelling the roots of one plant, exposed on a trench side, in order to distinguish them from the roots of neighbouring plants. Neilsen (1964) has described the use of ^{14}C-urea injection techniques for this purpose. In mixed stands Remezov, Rodin and Bazilevich (1963) suggest that the natural loss of roots each year may be roughly calculated by use of the formula,

$$\text{Root loss} = R = Wa + \frac{W - Wa}{n}$$

where 'Wa' is the weight of roots of the annual species, 'W' the total weight of roots present and 'n' the average number of years length of life of the perennial roots.

Pearson (1965) tackled the problem of a mixed stand in his studies of production in desert communities by giving production values based on each of three different assumptions as to average root age. Productivity of the desert communities, both roots and tops, was increased in grazed areas, and here annual root production was about 16% of the total root biomass. In the ungrazed areas root production was estimated as between 6% and 16% of root biomass depending on the assumption of root age.

Energy values

The energy relationships of root production have been very little studied apart from occasional references to the over-all energy value of roots (Hadley and Kiechkhefer, 1963). These workers obtained caloric values at monthly intervals and showed seasonal trends with peaks in the spring and autumn in the roots of prairie grasses. Roots of perennial woody species show marked changes in carbohydrate content throughout the season (Priestley, 1962), and caloric values may also be expected to vary between the different types of new growth, for instance, extension root tips and secondary vascular tissue.

Further research to improve estimates of root production

The almost complete lack of knowledge of root life-history and periodicity of root growth of many species is one of the main factors preventing advances in the planning and interpretation of biomass sampling.

Root observation techniques cannot contribute quantitatively to root production studies but the knowledge gained from wider application of the method would help immeasurably in biomass sampling. Observation studies would be valuable in estimating the average length of life of roots, the time taken for significant colour changes to occur, the estimation of root age and periodicity of root growth.

Root observation can also provide valuable information on the root feeding habits of the soil fauna. If it is one purpose of root production studies to estimate the amount of food available to root feeders, then it is important to know which parts of the root system serve as food for the species under investigation, the woody roots, the succulent new roots, the root hairs or the mucilage excreted by roots.

Once an estimate of biomass has been obtained the biggest unsolved problems in estimating root production are the natural loss of root material by death and decay, the depredations by plant and animal parasites and the loss of root cap cells, mucilage or exudation from healthy roots. The presence of mucilage layers around root surfaces and their colonization by bacteria has been demonstrated by Jenny and Grossenbacher (1962) with electron micrographs and the presence of very thick layers of mucilage on the roots of xerophytes has been demonstrated by Reynolds (1943). Samtsevich (1965) has attempted to estimate the amount of dry matter lost from the roots of several agricultural plants in the form of mucilage. The tips of wheat roots have particularly prominent mucilage caps, as shown in Fig. 3, and Samtsevich measured the diameter of these caps, the diameter of the root tips and thus calculated the volume of mucilage on each root tip. By multiplying by an estimate of the number of root tips per crop per hectare he estimated the dry weight of mucilage deposited in the soil during the development of a wheat crop to be equal to the dry weight of the grain taken from the field of wheat.

More sophisticated techniques are beginning to be used, on a laboratory scale, in quantitative investigations of root systems which may in future be developed to give more accurate information on losses of root material. For seedlings of cereal plants a method of ^{32}P injection has been developed which results in complete labelling of the root system (Racz, Rennie and Hutcheon, 1964; Rennie and Halstead, 1965). Removal of all the living roots of the experimental plant from the growth medium at a later date, to estimate root production, may then be corrected, for root material which has become separated due to natural death, predators, exudation, etc., by estimating the radioactivity remaining in the medium. ^{32}P may only be used for short-term experiments and a tracer with longer half-life would be preferable, but the workers involved fully realize the importance of obtaining uniform distribution of the tracer and the difficulties of achieving this.

Another approach to the problem of estimating losses from roots has been by the use of $^{14}CO_2$ feeding (Bartholomew and McDonald, 1966; Shamoot, McDonald and Bartholomew, in press). Plants of eleven annual and perennial herbaceous species were enclosed in gas-tight growth chambers and supplied with measured quantities of $^{14}CO_2$ in the atmosphere. In these experiments, some of which continued for 212 days, the calculated quantities of organic material left in the soil, after all living roots had been removed, ranged from 20·2 g to 49·4 g of organic debris per 100 g of roots removed, except for one abnormally low value. The authors state that the quantities of organic debris measured at the time of harvest underestimate the total

Figure 1. (a) Much-branched system of small apple rootlets, 28th June, 1965

Figure 1. (b) Same rootlets 12 months later. The cortex of all rootlets has turned dark brown but in most places has not rotted away

Figure 1. (c) Same area photographed 11 months later in May, 1967. The central root and many lateral rootlets have rotted away completely

(Engraved lines are approx. ½ in apart)

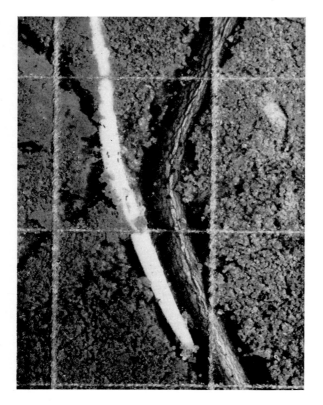

Figure 2. (a) New white extension root and brown thickened woody root of apple, 18th September, 1964. The woody root increased in diameter by 0·52 mm and 0·48 mm in 1963 and 1964, respectively

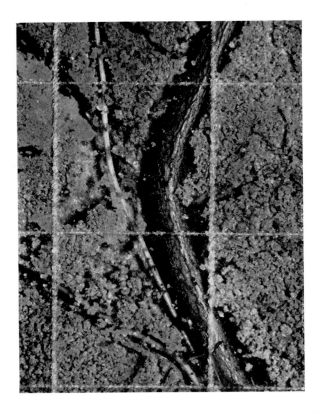

Figure 2. (b) The same area 2½ years later. Only the narrow central vascular cylinder of the extension root remains. The woody root has not thickened any more in 1965 and 1966
(Engraved lines are approx. ½ in apart)

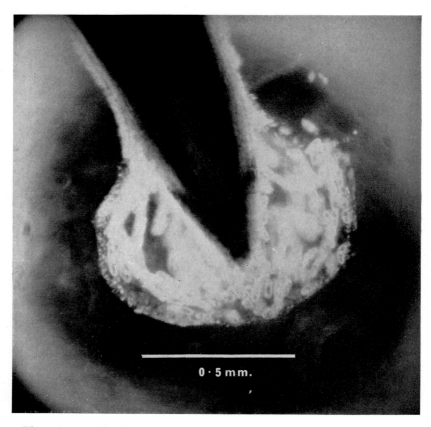

0·5mm.

Figure 3. Root tip of wheat root showing mucilage cap containing loose chains of cells (Photograph taken by Miss M.M. Fuller with dark field illumination)

deposition because much of the plant debris would have been extensively decomposed by that time and the ^{14}C released as $^{14}CO_2$ into the atmosphere.

Further refinement of this method would permit measurement of the CO_2 released by respiration in the soil but this would include the products of both root respiration and microbial respiration and it would then be desirable to determine these two components separately.

Discussion

An examination of methods that have been used to study root production leads to the conclusion that none of the methods at present available will give a very accurate measurement. In previous production studies roots have been included simply to complete the over-all picture and in these situations rough estimates of root production have been considered better than no estimate at all. If, however, the soil biologist wishes to use the data as a starting point for studies of energy flow through the food chains in the soil, rough estimates are not really good enough.

Work on roots is always difficult and laborious but at least the difficulties in obtaining an accurate estimate of biomass can be overcome by thorough planning of the sampling programme and the availability of sufficient labour. The problems of estimating the natural losses of root material between sampling dates, however, cannot be solved by labour alone. Considerable skill and ingenuity will be required and the examples quoted above perhaps point the way in which such studies might develop in the future.

Olsen (1963) has said that 'the kinetics of underground production and loss rates remain as one of the most challenging ecological and agricultural problems during the half-century to come' and the present survey underlines the truth of these words.

References

BARTHOLOMEW W.V. & McDONALD I. (1966). Measurement of the organic material deposited in soil during the growth of some crop plants. In: *The Use of Isotopes in Soil Organic Matter Studies.* FAO/IAEA Tech. Meeting, Brunswick-Völkenrode, 1963, 235–242.

BASKERVILLE G.L. (1965). Estimation of dry weight of tree components and total standing crop in conifer stands. *Ecology,* **46,** 867–869.

BRAY J.R. (1963). Root production and the estimation of net productivity. *Canad. J. Bot.* **41,** 65–72.

COKER E.G. (1958). The root development of black currants under straw mulch and clean cultivation. *J. hort. Sci.* **33**, 21–28.

DAHLMAN R.C. & KUCERA C.L. (1965). Root productivity and turnover in native prairie. *Ecology*, **46**, 84–89.

GREENLAND D. & KOVAL J. (1960). Nutrient content of the moist tropical forest of Ghana. *Plant and Soil*, **12**, 154–174.

HADLEY E.B. & KIECKHEFER B.J. (1963). Productivity of two prairie grasses in relation to fire frequency. *Ecology*, **44**, 389–395.

HEAD G.C. (1966). Estimating seasonal changes in the quantity of white unsuberized root on fruit trees. *J. hort. Sci.* **41**, 197–206.

HEAD G.C. (1968). Seasonal changes in the diameter of secondarily thickened roots of fruit trees in relation to growth of other parts of the tree. *J. hort. Sci.* **43**, 275–282.

JENNY H. & GROSSENBACHER K. (1962). Root–soil boundary zones as seen by the electron microscope. *Calif. Agric.* **16**, 7.

JUTRAS P.J. & TARJAN A.C. (1965). A novel hydraulic soil auger–screen shaker unit for the collection of soil samples. *Proc. Soil Sci. Soc. Fla.*, **24**, 154–158.

KOLESNIKOV V.A. (1962). *Root Systems of Top and Soft Fruit Plants and Methods of Studying Them.* [Russian.] Moscow, 190 pp.

MAGGS D.H. (1961). Changes in the amount and distribution of increment induced by contrasting watering, nitrogen, and environmental regimes. *Ann. Bot., Lond.*, N.S. **25**, 353–361.

MONK C.D. (1966). Root–shoot dry weights in loblolly pine. *Bot. Gaz.* **127**, 246–248.

NEILSEN J.A. JR. (1964). Autoradiography for studying individual root systems in mixed herbaceous stands. *Ecology*, **45**, 644–666.

ODUM E.P. & ODUM H.T. (1959). *Fundamentals of Ecology.* 2nd Ed. Philadelphia and London, Saunders.

OLSEN J.S. (1964). Gross and net production of terrestrial vegetation. *J. Ecol.* **52** (suppl.), 99–118.

ORLOV A.JA. (1955). The role of feeding roots of forest vegetation in enriching soils with organic matter. [Russian.] *Pocvovodenie* 1955, No. 6, 14–20.

PEARSON L.C. (1965). Primary production in grazed and ungrazed desert communities of eastern Idaho. *Ecology*, **46**, 278–285.

PRIESTLEY C.A. (1962). Carbohydrate resources within the perennial plant. *Tech. Commun. Bur. Hort., E. Malling*, No. 27.

RACZ G.J., RENNIE D.A. & HUTCHEON W.L. (1964). The ^{32}P injection method for studying the root system of wheat. *Canad. J. Soil Sci.* **44**, 100–108.

REMEZOV N.P. (1959). Methods of study of the biological turnover of elements in the forest. [Russian.] *Pocvovodenie* 1959, No. 1, 71–79.

REMEZOV N.P., RODIN L.E. & BAZILEVICH N.I. (1963). Instructions on methods of studying the biological cycle of ash elements and nitrogen in the above ground parts of plants in the main natural zones of the temperate belt. [Russian.] *Bot. Zhur.* **48**, No. 6, 869–877.

RENNIE D.A. & HALSTEAD E.H. (1965). A ^{32}P injection method for quantitative estimation of the distribution and extent of cereal grain roots. In: *Isotopes and Radiation in Soil–Plant Nutrition Studies.* Int. Atomic Energy Comm., Vienna, 1965. 489–504.

REYNOLDS M.E. (1943). *Anatomy of a Xerophyte: Dalea spinosa.* Dissertation. University of California at Los Angeles.

RODIN L.E. & BAZILEVICH N.I. (1965). *Dynamics of the Organic Matter and Biological Turnover of Ash Elements and Nitrogen in the Main Types of World Vegetation.* [Russian.] Moscow, Leningrad, Nauka.

ROGERS W.S. (1969). Amount of cortical and epidermal tissue shed from roots of apple. *J. hort. Sci.* **43**, 527–528.

ROGERS W.S. & VYVYAN M.C. (1934). Root studies. V. Rootstock and soil effect on apple root systems. *J. Pomol.* **12**, 110–150.

SAMTSEVICH S.A. (1965). Active excretions of plant roots and their significance. [Russian]. *Fiziol. Rast.* **12**, No. 5, 837–846.

SCHUURMAN J.J. & GOEDEWAAGEN M.A.J. (1965). *Methods for the Examination of Root Systems and Roots.* Wageningen, Centre for Agricultural Publications and Documentation.

SHAMOOT S., McDONALD I. & BARTHOLOMEW W.V. (in press). Rhizo-deposition of organic debris in soil. *Soil Sci. Soc. Amer. Proc.* **32**, 817–820.

TROUGHTON A. (1957). *The Underground Organs of Herbage Grasses.* Bull. No. 44, Commonwealth Bureau of Pastures and Field Crops, Hurley, Berkshire.

WEAVER J.E. & ZINK E. (1946). Length of life of roots of ten species of perennial range and pasture grasses. *Plant Physiol.* **21**, 201–217.

3

Plant Litter

A. MEDWECKA-KORNAŚ

Introduction

Above-ground litter produced by macrophytes—trees, shrubs, and herbaceous plants—plays a very important role in terrestrial ecosystems, especially in nutrient circulationd an energy transfer between plants and soil. The influence of litter upon the amount and chemical features of humus and on the micro-climate of the soil surface is also distinct; when accumulated into the deeper layers it forms a particular life environment and may be used by man for different economic purposes. Some of its components participate in the regeneration of plant communities.

For all these reasons litter, and especially forest litter, has already been the object of several studies (see classical data of Ebermayer, 1876; review by Lutz and Chandler, 1946; and Bray and Gorham, 1964), but methods for quantitative estimation are only described in a few publications (Chalupa, 1961; Korčagin, 1960). These quantitative methods are relatively simple, but have been developed in different ways by various authors. The aim of this paper is to propose ways by which such methods can be standardized to facilitate data comparison, but at the same time give scope for adaptation to different local conditions. Further experiments are still needed to define these methods precisely. The present contribution is based on literature available to the author, the IBP Methodology Leaflet by Ovington and Newbould, and experience obtained during investigations at the Nature Conservation Research Centre, Kraków (Medwecka-Kornas, 1967). The IBP Handbook No. 2 by Newbould (1967) was also taken into consideration.

Some of the more important definitions

The term litter should be used for all ecosystems (forests, grasslands, deserts) and should indicate all that material lying on the soil surface which is mainly

24

composed of dead plants or their shed organs. The 'standing' dead matter (e.g. tree stems which have not yet fallen) will not be included here.

The most important components of litter are leaves, budscales, twigs, bark, inflorescences, and fruits or seeds (small items), as well as branches and fallen stems (large items). The greater part of this material will sooner or later decay, with the exception of fruits which provide for the development of new seedlings.

The litter present at a given moment in a definite area of an ecosystem may be considered as its biomass or 'standing crop' and may be expressed in weight per unit area (e.g. in kg/ha or in energy units, kcal/ha).

The amount of litter formed and shed by the ecosystem within a definite period should be called litter production and expressed, possibly in relation to the whole year, e.g. as kg/ha/annum. In this definition the litter production will be the litter fall of the whole year, although not the increment of dead material, because part of the latter can be retained by living plants above the soil surface.

The decrease of the amount of litter in an ecosystem, caused by decay and mineralization, animal consumption, wind transport, harvest by man, and so on may be generally called its disappearance.

Litter accumulation depends on the rate of production, as well as on the rate of disappearance.

With regard to the age of litter and its degree of decay, the litter accumulated on the soil surface (A_0) may be divided into three layers: L—fresh organic material, F—partly decomposed litter, H—amorphous litter. These layers were previously designated A_0', A_0'', and A_0''', respectively.

Methods for the estimation of litter production

General principles. Plots for these investigations should be as uniform as possible and should represent one plant association, or even only one lower unit, a sub-association or facies. It seems reasonable to restrict the dimensions of such a study area to 1 ha (100×100 m), and when data on litter fall in a larger or more varied site are needed to divide the site into smaller parts which should be considered separately.

In litter sampling two main approaches may be recognized:

1. Continuous trapping of the material shed by an ecosystem with safeguards against its disappearance (litter trap method).

2. Single or repeated estimation of the amount of litter on the soil surface (ground litter sampling). This method may be used with some modifications for small and large items of litter. Also knowledge of the rate of disappearance may be needed in the case of small items.

The traps, as well as the small plots for litter sampling, may be distributed in the field at random or systematically. However the most appropriate method seems to be the distribution by 'Golden Points' (which can be found in the field with the help of special tables; Zubrzycka, 1960)*. This provides for a more even distribution than the random method and also allows the selection of additional points in the same area.

The number of samples (litter traps or plots for ground sampling) must be related to sample size, but for statistical reasons there should be a minimum of 25–30 per ha. In richer plant communities with a more variable spatial structure, or where the interest is in minor species (Newbould, 1967), the number of samples and then, if necessary, their size should be enlarged. It seems reasonable to suggest that the number and size of samples should be large enough to obtain an accuracy of 5% at 95% confidence limits (Ovington and Newbould, IBP Method. Leafl.).

Sample traps and plots should be circular to reduce the edge effect.

The frequency of litter collecting depends upon the phenology of the ecosystem and the method adopted. However the estimation should give results representative of the whole year or of the total growing season, and be based on several years (minimum 3 years) observations.

The collected material should be sorted into species and then into plant organs, dried at a temperature of 85°C–105°C to a constant dry weight, weighed or evaluated with regard to its caloric value, ash content, and so on. Production may then be calculated with the use of the appropriate formulae.

Litter trap method. A very large number of litter traps have been described in the literature: bags, buckets, boxes, shallow trays, and funnels (Newbould, 1967); only those which seem the most suitable will be mentioned here (Fig. 1). For litter trapping in forests, conical bags suspended from hoops about 1 m above the soil surface are highly recommended (Ovington and Newbould, IBP Method. Leafl.; Ovington and Murray, 1964; and Newbould, 1967). These bags should be freely permeable to water, e.g. made from nylon mesh, cheese cloth, or sail cloth, and have a minimum diameter of 0·5 m

* 'Tables of Golden Points' by Zubrzycka are available at the Nature Conservation Research Centre, Lubicz 46, Kraków.

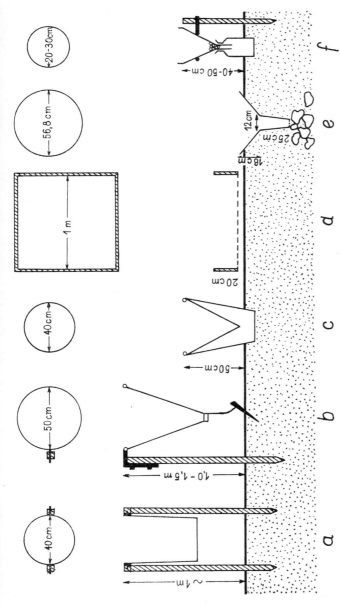

Figure 1. Some types of litter traps: a—simple bag of sail cloth used in the Ojców studies; b—bag of nylon mesh, cheese-cloth, sail cloth, etc., suspended from the hoop and pegged (may be weighed) to prevent it blowing inside out; c—some receptacle like a plastic dustbin or bucket with a bag of terylene gauze inside; d—shallow trays with wooden boards and terylene or nylon net at the bottom; e—funnel of tinplate with screen at lower end, buried in the soil; the stone layer promotes the water drainage; f—micro-litter trap consisting of polyethylene funnel with glass wool plug in lower part and outlet into polyethylene bottle. In upper part of the figure schemes of trap openings; below, their vertical cross-sections. For further explanations see: a—Myczkowski (in Medwecka-Kornaś, 1967); b, c, d, and f—Newbould (1967); and e—Korčagin (1960).

(1962 cm²) at the top, and a depth of 60–70 cm. They should be pegged or weighted to prevent litter blowing out of them.

Experiments made by Andrzejewski, Borowski and Olszewski (private communication) in the Białowieża Forest (North-Eastern Poland), in which a series of square bags of 25 different sizes were used, showed that the optimal size of the opening was approximately 32 × 32 cm (1000 cm²). In similar investigations near Kraców, bags with a round opening of 1250 cm² were used with good results. This order of dimensions seems therefore to be generally adequate for temperate forests.

For some types of communities, especially tropical forest, it may be necessary to use larger bags, but this has not yet been tested in practice (Kira *et al.*, 1967). Precautions must be taken to prevent the consumption of collected seeds by animals, e.g. by covering some of the traps with protective screens.

For sampling seeds and fruits, funnels of tinplate with screens or collecting chambers at their lower part are especially convenient (Newbould, 1967). When relatively flat and partly buried in the soil funnels can be used for collecting material from the ground flora; similarly one can adapt boxes, buried funnels and trays, to sample litter from the lower layers of ecosystems, which is not possible when using bags. The flat, shallow traps however, are useful only in special conditions, e.g. where dense vegetation prevents blowing off of the trapped material.

The trapped material should be removed frequently and at shorter time intervals in a warm and moist climate than in a cold and dry one. In temperate forests removal should take place once a week or at least once a month. The total of all the collected litter throughout the growing period may be used to compute production in relation to the sampled area.

Pieces larger than half the trap diameter should be collected separately using wooden frames with punctured bottoms or directly from the ground (see below) and their mass included in the results. Very small items of litter, e.g. plant pollen, may be sampled using sticky paper or bottles with funnels (Newbould, 1967).

Ground litter sampling. Several methods can be used here: hand sampling, raking, mechanical sampling and may be utilized on different kinds of plots. In the case of small items, the size and number of plots should be chosen in the same way as in the trap method. The author's experiences show that in temperate deciduous forest satisfactory sampling of ground litter may be achieved from 25 (30) plots of about 1000 cm² each, or, if the litter layer is homogeneous, from about half this area. Relatively small samples also suffice in turf communities, e.g. meadows (Jankowska, 1967).

Plots laid out in the field should be encircled by hoops (or frames) fixed in the soil. The litter contained within may be cut along the frame by a knife and lifted out. In this way the litter originating from all layers of the ecosystem, such as tree crowns, undergrowth, and ground flora is included in the investigation and the whole 'standing crop' is easy to estimate. If it is possible to distinguish the litter of the current season, a general evaluation of production for this period may also be done (in the separation of new litter fall from older layers the use of a screen, or previous clearing of sampled plots, is very helpful). This type of estimation usually results in an underestimation of the production values, especially if some components of the litter disappear relatively quickly.

Litter production (P) may also be detected by two or more samplings of the ground layer during the vegetational season and the subsequent calculation of differences in quantity. Then, for more exact results, the process of disappearance must also be taken into account according to the formula:

$$P = D + S_1 - S_0$$

Where D =amount of litter which disappeared during the time interval, S_0 =standing crop of litter at the beginning of the time interval, and S_1 = standing crop of litter at the end of the time interval.

To obtain information on disappearance special methods should be used. Evans and Wiegert (1964) in their study on primary production in a grassland community applied a paired plots method. At the beginning of the time interval the dead material from the first plot was collected and weighed, and the green material was removed from the second. This prevented the increase of dead material on the second plot. At the end of the time interval the dead material from the second plot was collected and weighed. By comparing the amounts of dead material collected at different times in different plots, the disappearance (D) can be calculated. This method, also used with good results by Lomnicki and Bandoła (in press) and modified, is especially appropriate for closed, herbaceous vegetation.

The disappearance (D) can also be found by placing samples of litter in mesh nets on the ground of the ecosystem, and weighing them at the beginning and end of the experiment.

The production of large items of litter may be evaluated by counting, measuring and sampling at intervals of time from definite areas, e.g. 10×10 m

Chapter 3

or narrower rectangular plots, e.g. 1×5 m (Sukačev and Dylis, 1966). For isolated trees triangular plots covering about 10% of the area shaded by the crown are also recommended (Korčagin, 1960).

To avoid the difficulties of statistical computation connected with the distribution of samples and the definition of their dimensions, Kira *et al.* (1967) used a long belt of plastic net which they suspended immediately above the ground.

Litter production is sometimes also evaluated *indirectly* from the annual increment of living green organs, e.g. by collecting leaves from cut trees or shrubs and taking into account their life period (Müller and Nielsen, 1965). It is, however, necessary to stress here that there may arise a difference between the biomass of produced leaves and the amount of shed litter caused by animal consumption of living plant parts, e.g. by caterpillars of insects (Carlisle, Brown and White, 1966).

Some features of litter production in different ecosystems—conclusions

Litter production is proportionate to the biomass of vegetation, being usually equal to or smaller than the annual increment of the latter. However, litter production attains higher values in favourable habitats (e.g. in moist tropical climates) than in the less favourable environments of cold or dry zones. At the same time the accumulation of dead organic material is very uneven (Kira and Shidei, 1967), e.g. it is higher in shrubby tundra and boreal forest (taiga) than in tropical forests. The litter fall itself also shows great variability in its rhythm and may be continuous or seasonal (Bray and Gorham, 1964; Bazilevič and Rodin, 1965). All these facts are very important in detailed litter investigations in different zones and ecosystems.

In tropical evergreen forest litter fall may be estimated daily for selected periods of the year (Kira *et al.*, 1967). However, this will not detect any rhythm in litter fall, which may well exist (Richards, 1966). In evergreen coniferous forest continuous sampling is called for and traps may be similar to those used for fruit-crop measurements (Tolskij, 1932; —cf. Korčagin, 1960). In this situation leaves are retained for several years and fall gradually; furthermore the litter layer is sometimes difficult to separate from raw humus (mor). In seasonal deciduous forests a single sample taken just after litter fall may provide general information about leaf production. It is more desirable, however, to trap litter throughout the whole vegetative season, thereby avoiding underestimation (Ovington, 1963). In more open forest and savannas litter distribution becomes less regular and, during collecting, special precautions should be taken against the effects of wind.

Litter production of grassland depends to a high degree upon grazing and mowing, both of which should be taken into consideration in the calculations. In steppes, the dead material forms the layer of 'mat' (Sukačev and Dylis, 1966) and here sampling from the ground surface seems to be most adequate. Ground litter sampling is also convenient in heathlands (Cormack and Gimingham, 1964). In low arctic and alpine communities dead material may be collected in the same way as for the estimation of plant increment (Bliss, 1962) or in the biomass studies of Aleksandrova dated 1958 (cf. Sukačev and Dylis, 1966). On bogs the dead matter is stored for several years and its production can be measured by the annual increase in peat (Lieth, 1967). In deserts, the litter production of perennial plants may be estimated by sampling dead material and seeds from selected plots (Sukačev and Dylis, 1966), and that of ephemerals after their maximal biomass development.

It is evident from the few selected examples that in some cases (depending mainly on the structure of the vegetation) trapping may be more useful whilst in others ground sampling methods are better. Usually the best result can be obtained by joint use of both these approaches. For the comparison of litter production in different ecosystems it is not only important to use satisfactory methods for its estimation, but also to obtain descriptive information about local factors influencing production, such as exposure, soil moisture, tree density, human management, and so on. Without them a critical discussion of results, and detection of differences of real geographical value will not be possible.

Acknowledgements

The author is indebted to the participants in the Paris Symposium for their remarks, and particularly to Professors A.A. Korčagin, W. Kühnelt, A. Macfadyen and P.J. Newbould for reading the text and making helpful suggestions.

References

Bray J.R. & Gorham E. (1964). Litter production in forests of the world. In: *Advances in Ecological Research*. J.B. Cragg (ed.). London and New York. **2**, 101–157.

CARLISLE A., BROWN A.H.F. & WHITE E.J. (1966), The litter fall, leaf production and the effects of defoliation by *Tortrix viridana* in a sessile oak (*Quercus petrea*) woodland. *Ecology*, **54**, 1, 65–85.

CHALUPA V (1961). Contribution to the leaf production of the beech and oak forest. *Reports of the Forest Research Institutes of Czechoslovakia*, **23**, 35–62.

CORMACK E. & GIMINGHAM C.H. (1964). Litter production in *Calluna vulgaris* (L.). *Ecology*, **52**, 2, 285–297.

EBERMAYER E. (1876). *Die gesamte Lehre der Waldstreu mit Rücksicht auf die chemische Statik des Waldbaues.* Berlin, Julius Springer.

KIRA T., OGAWA H., YODA K. & OGINO K. (1967). Comparative ecological studies on three main types of forest vegetation in Thailand. IV. Dry matter production, with special reference to the Khao Chong rain forest. *Nature and Life in Southeast Asia*, V, 149–174.

KIRA T. & SHIDEI T. (1967). Primary production and turnover of organic matter in different forest ecosystems of the western Pacific. *Japanese Journal of Ecology*, **17**, 3, 70–87.

KORČAGIN A.A. (1960). Methods of determination of the seed productivity of forest trees and forest communities. In: *Field Geobotany.* II. E.M. Lavrenko & A.A. Korcagin (eds.). Academy of Sciences of the USSR Press, Moscow-Leningrad. 41–132. (In Russian.)

LIETH H. (1965). Indirect methods of measurement of dry matter production. In: *Proceedings of the Montpellier Symposium*, F.E. Eckardt (ed.). Methodology of plant ecophysiology. Paris (Arid Zone Research, XXV), 513–518.

LUTZ H.J. & CHANDLER R.F. JR. (1946). *Forest Soils.* New York, John Wiley.

LOMNICKI A., BANDOLA E. & JANKOWSKA K. (1968). Modification of the Evans-Wiegert method for the estimation of net primary production. *Ecology*, **49**, 1, 147–149.

MEDWECKA-KORNAŚ A. (1967). Ecosystem studies in a beech forest and meadow in the Ojcow National Park. In: *Studia Naturae, ser. A.* A. Medwecka-Kornas (ed.). Kraków, Wydawnictwo Naukowe, Zaklad Ochrony Przyrody. 213.

MÜLLER D. & NIELSEN J. (1965). Production brutto, pertes par respiration et production netto dans la forêt ombrophile tropicale. *Det Forstlige Forsøgsvaesen i Danmark*, **29**, 69–160.

NEWBOULD P.J. (1967). Methods for estimating the primary production of forests. *IBP Handbook No.* 2. Oxford and Edinburgh.

OVINGTON J.D. (1963). Flower and seed production. A source of error in estimating woodland production, energy flow and mineral cycling. *Oikos*, **14**, fasc. 2, 148–153.

OVINGTON J.D. & NEWBOULD P.J. General procedures for determining the organic. production of woodlands. *IBP Methodology Leaflet*, mimeogr. by IBP Central Office.

RICHARDS P.W. (1957). *The Tropical Rain Forest. An Ecological Study.* Cambridge, The University Press. 193–198.

RODIN L.E. & BAZILEVIČ M I. (1964). The biological productivity of the main vegetation types in the northern hemisphere of the Old World. *Doklady Akademii Nauk SSSR*, **157** (1). 215–218. (In Russian.) and (1966) *Forestry Abstracts*, **27**, No. 3, 369–372.

SUKAČEV W. & DYLIS N. (1966). *Programme and Methods of Biogeocoenological Investigations.* Moscow, Publishing Office 'Nauka'. Academy of Sciences of the USSR. (In Russian.)

Wiegert R.G. & Evans F.C. (1964). Primary production and the disappearance of dead vegetation on an old field in South-eastern Michigan. *Ecology*, **45**, 1, 49–63.

Zubrzycka L. (1960). On the distribution of sampling points in a plane. *Zastosowania Matematyki*, **V**, (1960), 161–171.

4

Heterotrophic Microflora

D. PARKINSON, T.R.G. GRAY, J. HOLDING
and H.M. NAGEL-DE-BOOIS

I Introduction

The following contribution is the result of detailed discussions held by the
soil microbiologists present at the UNESCO–IBP Symposium held in Paris
(1967), thus numerous workers (other than those listed above) have presented
ideas and details included herein.

Over the past two decades a large number of techniques and modifications
of techniques have been devised for the qualitative and quantitative study of
soil microorganisms and a good deal has been written regarding the relative
merits and applicability of these techniques. It is not the purpose of this
contribution to survey the whole range of methodology, but rather to
concentrate on the basic measurements recommended for use in IBP PT
projects:

    'a. A count of the fungi, bacteria and actinomycetes, as determined by a
        dilution plate estimation.

    b. A direct measure of the number of bacteria and the amount of fungal
       hyphae present.'

    (IBP News No. 2, p. 16.)

II The dilution plate count technique

A Reasons for using the dilution plate count

A major aim of the International Biological Programme is to obtain figures
for rates of production of living material throughout the world under
different environmental conditions. At present, no techniques are available
to determine the rate of microbial production in soil and as an interim
measure, it has been suggested that microbiologists at least supply biomass or
standing crop figures for different groups of microorganisms in the major
microhabitats in the soils under study. One way of doing this is to count the
microorganisms present in any one soil at a given time and multiply the
figure obtained by the average weight of one organism. The accuracy of this

34

biomass determination depends on the accuracy of the count and the average weight figure used. Figures for the average weight of single bacterial cells may be obtained from the literature, e.g. Alexander (1961) or determined for the population in question by measuring cell volume and multiplying this figure by cell density constants.

It is generally considered that the dilution plate count technique is useful for estimating the numbers of viable microbial cells in soil. It is also the most widely used technique for isolating microorganisms used in qualitative studies of the soil microflora. The technique is based upon the assumptions that when soil is dispersed in a liquid, the cells become detached from the soil and that when this cell suspension is mixed with an agar medium in a petri dish, each cell will grow and give rise to discrete colonies. By counting these colonies, the number of cells in the suspension can be determined and hence the number of cells in the original soil sample estimated.

These assumptions have been widely challenged, for not all cells are released from soil particles and, even when they are, they may still exist as small clumps of cells, rather than as individual cells. Not all the cells will revive and grow on any one agar medium, so dilution plate counts underestimate, perhaps seriously, the total viable population of cells. These and other errors associated with the technique are discussed by Skinner, Jones and Mollison (1952).

B Standardization of the plate count method

The most accurate biomass determinations will be made from counts which are closest to the actual number of cells present in the soil. Consequently, the highest viable count should be aimed for by varying the technique to suit the particular microorganisms and soils being examined. There are probably as many variations of the plate count technique as there are workers using it, variations often slavishly adhered to by their proponents without reference to improvements published over the years in the literature. It is almost possible to discover where a soil microbiologist graduated by watching him carry out a plate count. If comparable data are to be obtained from different sites throughout the world, it might be thought desirable to elect a standard method of performing the technique. However, it seems more practicable to use *attainment of the highest count possible* as the main standard, allowing individuals to achieve this within the limitations of their own laboratories. There are many reasons for rejecting a rigid standard method. Egdell *et al.*

(1960) showed very clearly that even when the same method was used on the same soil, different workers in different laboratories could not achieve uniformity. Furthermore, methods that provide high counts for one soil, are poor when used with other soils. Jensen (1968) points out that apparently minor changes in techniques are responsible for variation in results. Such findings suggest that when the plate count is used, it should be clearly and precisely described. This is especially true when results from an international study may be used by others to make generalizations about differences between soils.

The plate count technique consists of a large number of variable steps. Below are listed those steps thought to be most critical by the group of microbiologists at the UNESCO/IBP meeting (Methods of Study in Soil Ecology, Paris, 1967). Specific recommendations are made where general agreement was possible. Where such recommendations are not made, it is the responsibility of individual workers to select a technique and to describe it in such a way that the data can be adequately assessed. Ideally, variations of the technique should be selected only after they have been tested on any soil. If such tests are carried out, their results should be available for consultation. However, it is well known that testing all variations of the method is time consuming and since biomass determinations are not the main aim of the International Biological Programme, it is suggested that workers do not become preoccupied with plate counting at the expense of other work. Time would be better spent devising new approaches to measuring microbial productivity, whilst providing biomass measures with well described, if not standard and 'perfect' techniques.

C   Application of the plate count to different microbial groups

One other general point concerning the technique deserves mentioning. Data are only valid for the purpose of biomass determination if the organisms being counted are of approximately uniform size. It may be assumed that this is roughly so for bacteria, but it is not true for fungi or actinomycetes. Organisms in both these groups exist as differing lengths of mycelium and individual spores. It is recommended that estimates of fungal biomass be based on direct observation records of mycelial length, never on plate counts. If direct measurements can be made of actinomycete mycelium these should be used to calculate biomass, but in the absence of such data, plate counts may be used, not so much because they will give accurate data but because we know very little about the ecology of actinomycetes and the more information we can obtain the better.

Plate counts may be used for estimating yeast biomass. This group has been neglected hitherto and mycologists should consider including these organisms in their programmes.

D Critical steps in the dilution plate technique

1 *Preparation of soil suspensions*

The following variables should be described: (a) amount of soil used, (b) ratio of soil: diluent, (c) chemical nature and concentration of the diluent, (d) method of dispersal of soil in the diluent.

A full description of the dispersal method is essential since so many techniques have been described. It is not possible to recommend any specific technique, though many workers are now agreed that better dispersal is achieved by inserting a mechanical dispersal probe into the soil suspension, rather than mere shaking of the suspension. Examples of such techniques are the use of Waring blenders (Jensen, 1968), magnetic stirring bars (Hill and Gray, 1966) and trituration (Pochon, 1954). Whichever technique is adopted, both the speed and duration of dispersal should be given, and in the case of shaking methods, the mode of action and distance of throw of the shaker, e.g. reciprocal shaker, 200 throws per minute, 10 cm.

Soil should not be sieved before counting, except when comparing different plate count techniques. If soil is extremely heterogeneous and contains well-defined microhabitats, e.g. roots, leaves etc., plate counts may be referred to these habitats rather than to the soil as a whole. An estimate of the abundance of these microhabitats should then be included.

2 *Dilution of the soil suspension*

The following steps must be described: (a) time elapsing between dispersion and dilution of specimens, thus indicating if coarse soil particles have been allowed to settle out or not, (b) chemical nature and concentration of the diluent, (c) ratio of the dilutions, i.e. two fold, five fold or ten fold, (d) method used to make dilutions, i.e. pipetting or other method.

If pipettes are used, 1 ml blow-out pipettes are preferred. The accuracy of the pipettes should be known so that the pipetting error can be used when calculating the experimental error. Manufacturers are usually willing to provide such information for any particular batch of pipettes.

Pipetting steps should be as few in number as possible, and the procedure used described fully. Many workers prefer to fill and empty pipettes 10—12 times before transferring suspensions so that the adsorption sites on the pipette walls are fully saturated. This minimizes errors due to variations in the degree of bacterial adsorption.

3   *Preparation of petri dishes*

State the following in your description: (a) diameter of petri dish in centimetres, (b) volume and temperature of the agar used, (c) methods used to control or eliminate spreading bacteria, (d) method of medium inoculation, i.e. pour plates or surface inoculated plates.

A useful way to control spreading bacteria is to pour a thin layer of tap water agar into the dish (c. 10 ml) before addition of the agar nutrient medium (Orcutt, 1940). This eliminates spreading of organisms between the agar and the glass bottom of the dish.

4   *Composition of the medium (for heterotrophic organisms)*

In the matter of choice of medium, it is desirable to test a few broad groups of media before choosing one to use routinely. Any comparison should include at least one representative of a soil extract based medium, a medium containing peptone or other similar nitrogen sources and a synthetic mineral medium. The origin of all organic chemicals should be specified. The pH of the medium after sterilization, together with the sterilization procedure should both be quoted.

It is firmly recommended that selective inhibitory chemicals and antibiotics should be included in media. Media for the isolation of bacteria should include antifungal antibiotics and media for isolation of actinomycetes should contain antifungal antibiotics and antibacterial antibiotics, if bacteria make counting difficult. (N.B. Antibacterial substances may reduce the numbers of actinomycetes.) Yeast media should contain antibacterial antibiotics. Discussions of the problems involved may be found in papers by Johnson (1957), Martin (1950) and Williams and Davis (1965).

5   *Incubation of inoculated plates*

Plates should be incubated for at least 14 days and the count obtained after 14 days recorded. Counts after other intervals of time may be made at the discretion of the investigator. The temperature of incubation should be given.

This will normally be a temperature conducive to the development of meso-philes, e.g. 25°. Lower temperatures may be difficult to achieve without expensive cooling apparatus, both in the tropics and in centrally heated laboratories. In special cases, e.g. in tropical and arctic areas, where themo-philes and psychrophiles may be abundant, additional experiments may be required.

Plates should be incubated in an inverted position. If plastic disposable dishes are used, care should be taken to see that the agar does not dry out during incubation.

6 Recording the results

Colonies should be counted on plates at those dilution levels at which 20–200 colonies develop, on each plate. Wherever possible, the plates counted in any one experiment should be at the same dilution level. If fungi or spreading bacteria occupy more than about 15% of the surface of the agar, these plates should be rejected and not included in the count. Readily recognizable colonies of actinomycetes should not be included in bacterial plate counts and vice versa. If colonies are counted with the aid of a magnifying lens, details of the magnification achieved should be recorded. It is preferable that 5–10 replicate plates be prepared at each dilution. More replicates are unnecessary; it is possible to use less, but not desirable.

The degree of replication must be recorded, including the number of replicate dilution series made and, in the case of counts of organisms in composite soil samples, the number of sub samples used to make up the composite sample.

7 Expression of results

Counts should be recorded, but it is also necessary to convert these figures to biomass measurements from estimates of individual cell weights, obtained from determinations of cell size and density. Sufficient information concern-ing the soil should be presented so that workers in any part of the world can express their biomass figures per gram of oven dried soil, per cm^3 of soil under natural conditions and per sq. cm. of internal surface area of the soil and, if possible, per $metre^2$.

Results should be analyzed statistically to show the degree of error result-ing from the apparatus and experimental procedure used.

In order to determine the oven dry weight of a soil, sub samples of the soil used to make the soil suspension should be dried at 105° for 24 hours. Soils should then be cooled in a dessicator and weighed, taking great care to avoid uptake of moisture from the air during the weighing process.

III   Direct counts of microorganisms

A   Methods for estimating numbers of bacteria and fungi

Comparing the available techniques for direct observations of bacteria and fungi in soil it has been shown that for quantitative estimations the modified impression slide technique (Brown, 1958) and the soil sectioning technique (Burges and Nicholas, 1961) are not suitable for bacterial counts. Also for mycelial counts (measurements) there are several disadvantages in these techniques. The remaining available techniques are the agar film technique (Jones and Mollison, 1948) and the soil suspension slide technique (Witkamp, 1960). The agar film technique has been indicated to be the best method for quantitative assessments of mycelium in the soil (Nicholas and Parkinson, 1967). The soil suspension slide technique is particularly valuable for bacteria.

Measuring mycelium in litter samples has been attempted following the techniques of Minderman (1956) and of Hering and Nicholson (1964). But, during attempts to develop uniform techniques, it has been found that the Jones and Mollison method can be used for litter samples after drying them at 75°C and grinding (unpublished data T.F. Hering), or after homogenization without drying of the litter (unpublished data Elisabeth Jansen). Drying of litter may cause a reduction in the counts of bacteria and fungi and therefore should be avoided if possible.

Because the Jones and Mollison slides can be prepared for all soil layers and can be used for mycelial measurement as well as for bacterial counts it is recommended that this technique be used. If the Witkamp method is adopted, the counts should be compared with counts obtained using the Jones and Mollison technique so that cross-reference can be made. This is necessary as is shown by Nicholas and Parkinson (1967).

B   Agar film technique (Jones and Mollison, 1948)

1   *Short description of original method*

A soil sample is sieved. A known weight of soil (the weight used depends on the degree of microbial development in the soil and must be ascertained by preliminary tests) is ground up with 5 ml distilled water. This suspension, without the heavier sand fraction, is made up to 50 ml with 1·5% agar. From this suspension, kept at 40°C to prevent setting of the agar, 0·1 mm thick films are made on a haemocytometer slide. The agar films are allowed to dry on an ordinary microscope slide. The dried films are stained in phenolic analine blue and mounted in euparal. The number of bacteria or the length of mycelium per gram soil can be calculated by observation of 20 microscope fields per film on 4 replicate films.

(For description and statistical note see the original paper. For comparison of this method and the plate-counting technique see Skinner, Jones and Mollison (1952).)

2 *Additions to the technique*

(a) Preparations of leaf litter are made by grinding or homogenization of 2 gram of discs of several leaves, dry or in 5 ml water. In both cases the suspension is made up to 50 ml with 1·5% agar. It may be necessary to bleach the litter. The less drastic method of Hering and Nicholson (1964) seems to be better for the small leaf fragments than the stronger bleaching of Minderman (1956). Preparation and staining is the same as in soil preparations.

(b) For fungal counts chlorazol black E (British Drug Houses Ltd.) is recommended as a dye. Hyalin hyphae are selectively stained by this chitin-dye. 2 gram chlorazol black is dissolved in 1 litre 70% alcohol; filtration is necessary. Agar films are stained for 1 hour.

(c) The value of stains to distinguish between live and dead bacteria or mycelium has not been proved. (Acridin-orange staining may cause live bacteria to fluoresce green in the soil preparations.) In Jones and Mollison slides acridin-orange cannot be used to distinguish between live and dead parts of mycelium probably because the mycelial fragments are too short, due to grinding, to contain protoplasm. Jones and Mollison (1948) found a correlation of intensity of staining with viability of mycelium when using phenolic aniline blue. They also mention a variation of staining intensity for bacteria, but this has not yet been correlated with viability.

However, it is felt that in some way a crude index of the proportion of live and dead material should be made.

(d) It is impossible to lay down absolute instructions for the use of Jones and Mollison preparations for all litter and soil layers because of natural variations in microbial content. It is, therefore, recommended that the general method will be used, but that modifications (based on preliminary experimentation) should be used to enable the maximum count to be obtained. In all cases where data is published the method should be described in detail.

C   Estimation of production

It is generally felt that estimates of numbers at a given time are of very limited value, and that their importance has been over-emphasized. It is much more important to obtain some measure of production or turnover and to measure nutrient transfer. However, studies of nutrient transfer require many new studies which cannot be completed within the life-span of the IBP. At present there is no method available for the estimation of production and turnover in microbial populations, but at least three methods can be used in an attempt to obtain data on growth and death rates of microorganisms in soil.

1. The growth rate of fungi in soil and litter layers may be measured using the nylon net technique of Waid and Woodman (1957). Nylon gauze, which is biologically inert, is buried vertically in the soil and exposed for a defined time. After careful removal of the gauze from the soil the presence or absence of fungal hyphae is recorded in 200 meshes of each of 3 replicate nylon nets. This technique is used mainly to compare levels of hyphal activity in different soils. Nagel-de-Boois and Jansen (1967) used this method in combination with Jones and Mollison counts for estimating relative breakdown of mycelium in soils in different seasons.

Several factors operate in the decomposition of fungi in soil: direct activity of bacteria, enzymes, autolysis. Over one year there is an equilibrium between growth and decomposition. But these two phenomena, the interaction of which result in the mycelium-content of a soil at any one time, change independently in the course of a year. When the duration of hyphae in soil and the hyphal content of the soil is known for every month, the total mycelium turnover can be calculated for a whole year. Of course there are fungal species which are very resistant to decay, while other species are more susceptible to attack. For the IBP purposes, however, it is not necessary to deal with single species.

The estimation of the length of time an average hypha persists in soil can be made using the nylon net method. When a large number of pieces of nylon net are buried in the soil, effectively continuous observation can be made. Twice weekly three replicate nets are removed and observed, the percentage of meshes containing hyphae being recorded. After some days or weeks a time will be reached when the percentage of meshes with hyphae no longer increases because the decomposition of the first grown hyphae has started (nets removed from soil cannot be replaced because of the disturbance of the mycelium and the surrounding soil).

For examination of the same net on a number of occasions it is necessary to bury the nylon net behind the glass wall of a box with open top and bottom. The box can be filled with the soil of the profile. Sequential observations of the nylon net can then be made without disturbance of the net. A disadvantage of this is that there is no opportunity for the mycelium to cross the net in a natural way and there will be more water in the meshes due to condensation on the glass wall of the observation box.

For uniformity of observations it is desirable to use one kind of nylon gauze (net). A recommended type is single-threaded gauze with an open space of 40 µ (for example: Monodur 40, Vereinigte Seiden Webereien A.G., Krefeld, Germany). Re-use of nylon nets is possible after cleaning in KOH 50% for some days.

2. The pedoscope was developed by Perfil'ev and Gabe (Gabe, 1961). This apparatus is valuable for direct and continuous observations of micro-organisms, as shown by Aristovskaya and Parinkina (1962). The pedoscope can be removed from the soil for observation and then replaced without great disturbance. It allows analysis of the individual components of the total microflora and their relationships.

This method may be criticized because a non-natural substratum is provided, i.e., agar containing organic mineral complexes of humic acids. It is impossible to make quantitative estimations with this method.

3. Use of fluorescent brighteners which can be transmitted from parent to progeny is being investigated. This can be valuable especially in conjunction with incident fluorescent microscopy on leaf surfaces. This method has not yet been fully developed.

D Estimation of biomass

The conversion of numbers of bacteria and length of mycelium (obtained from Jones and Mollison preparations) to a common unit of weight should be made wherever possible. The size of bacterial cells and the diameters of fungal mycelium should be measured to allow conversion to volume and wet weight (assuming specific gravity of one). Cells and hyphae on Jones and Mollison slides will have changed in size (sometimes to 2/3 natural size) because of shrinkage in fixing and staining. Therefore it is best to measure the size of cells and hyphae on a few fresh preparations, and by comparison with Jones and Mollison preparations obtain a correction factor for shrinkage. It would be desirable to separate weight of dead and live organisms, but this is not yet feasible. It must be realized that in this estimation of biomass mycorrhizal fungi are not included.

E   Expression of results

For comparison of results it is necessary wherever possible to estimate the numbers of microorganisms in the main components or horizons of the soil, i.e. litter, humus, mineral soil and root surface fractions. The proportions of these fractions in the soil should be measured so that the numbers can be expressed per gram (oven dry organic matter and oven dry soil), per cm^3 of soil or organic matter under natural conditions, per sq. cm of internal surface area, and per metre^2.

IV   Techniques for qualitative studies on soil fungi

At this juncture, some comments should be made on methods for studying the qualitative nature of fungal populations in soil. This special treatment of soil fungi can be made, because qualitative studies on soil actinomycetes and soil bacteria are normally based on the application of the soil dilution plate technique which has been described in detail above. This technique is only of value in studies of soil fungi if data on the nature and quantity of fungal spores in soil samples are required (a case for the use of the soil dilution plate technique in soil fungus studies was made by Montégut, 1960).

In assessing the nature of fungal populations of soil probably the simplest and quickest technique available is the soil plate technique (Warcup, 1950; 1951). This technique involves the use of no complex apparatus and, despite its shortcomings (in its application to soil fungus studies it suffers from all the shortcomings of the soil dilution plate technique), it is useful for preliminary studies on the nature of soil mycofloras (where a distinction between the fungi present as spores and those present as hyphae is not required). However, if an attempt at such a distinction is required then other techniques must be used and washing techniques appear particularly useful in this respect.

The soil washing technique (Williams, Parkinson and Burges, 1965) or some modification (Hering, 1966; Gams and Domsch, 1967) has been used for various ecological investigations. It is particularly useful (and efficient) when applied to sandy soils. As well as allowing the removal of many of the fungal spores in soil samples, this technique allows some separation of soil microhabitats. If this technique is to be used it is essential that it be tested to ascertain the appropriate number of washings required for maximum removal of fungal spores from the soil under study.

Harley and Waid (1955) pointed out the value of washing techniques in studies on fungi colonizing various plant parts. Their technique is very valuable for studying fungi colonizing leaf litter in its various stages of decomposition, and for studying fungi colonizing living or dead plant roots. In other words, this technique is particularly valuable in the study of discrete, manipulable microhabitats (or groups of microhabitats).

The choice of the nutrient medium onto which washed material (from whose surfaces all free water has been removed) is plated is an important factor in determining the fungal species isolated. This choice must depend on the group or groups of fungi under study. Media such as Czapek-Dox with or without added yeast extract are frequently used to isolate a broad (not complete) spectrum of fungal species. The addition of selective antibiotics, a frequent practice, must be made with the care referred to in an earlier section of this contribution.

Study of specific physiological groups of soil fungi is frequently initiated by attempts at isolation of representatives of the groups required using selective isolation media (e.g. Eggins and Pugh, 1962; Gray and Bell, 1963; Mathur and Paul, 1967). Another approach is by 'baiting' soil with specific substrates (e.g. cellulose strips—Tribe, 1957; chitin strips—Gray and Bell, 1963), for known periods of time, then isolating the fungi colonizing the 'bait' (which can be done using some form of the washing technique).

Temperature of incubation of plates prepared for fungal isolations may play an important role in determining the spectrum of species isolated. Most published data is based on fungal isolations on plates incubated at room temperature or at 25°, thus the mesophilic component is selected. Where studies are directed onto natural situations where extremes of temperature are experienced, it is essential to supplement the normal method of incubation in order to discover what species (if any) are capable of activity under naturally occurring temperature regimes.

In general it can be said that, unless the isolation plates are overrun with fast-growing potentially antagonistic species (e.g. *Trichoderma viride*), the plates should be observed over as long a period as is possible.

The foregoing comments are by no means exhaustive because the consideration of the detailed species structure of soil mycofloras and successional sequences on organic matter is not of prime importance to the basic aims of IBP (where analysis of productivity is the goal). However, such considerations are important if real understanding of the biological interactions in the decomposer cycle are required, or where detailed information on fungal activity in soil microhabitats is desired.

Many of the points raised in this section have been dealt with extensively in several review articles (e.g. Warcup, 1960, 1967; Garrett, 1963; Parkinson, 1967; F.A.O. Soils Bulletin No. 7, 1967); these contain comprehensive reference lists.

V   Soil sampling for microbiological analyses

Whenever possible representative soil samples should be obtained by the standard procedure outlined below. Microbiological data from different centres can then be more accurately and reliably compared.

Before sampling the site, any available data on vegetation, including the occurrence of fungal fruiting structures, leaf litter and roots, should be examined. This information together with any details of the physical and chemical characteristics of the soil (e.g. water status, particularly the level of the water table, pH, temperature, percentage organic matter) will assist the investigator in determining the number of different types of habitat within the site.

The object of sampling is to obtain a soil core which is disturbed as little as possible. The equipment recommended comprises a strong metallic cylindrical corer, open at both ends, and with a horizontal handle near to one end. A clamping device running along the length of the corer enables the soil sample to be readily recovered. An alternative device useful for obtaining only shallow soil samples is a cylindrical metallic food container, e.g. fruit tin. The soil samples are obtained by pressing the cylinders vertically into the soil. Soil sampling equipment should be as clean as possible and not contaminated with visible quantities of soil. No data are available on the most suitable diameter for the cylinders. In hard mineral soils, a strong corer of about 5 cm in diameter is suitable, whereas in highly organic soils, cylinders of 10–15 cm are in routine use. The length of corer depends on the depth of sample required, and especially the location of particular horizons. In some arable soils, a depth of only 10 cm might be required but samples of 60 cm depth have been obtained satisfactorily in peat soils.

If a microbiological analysis of a number of samples is undertaken, the results can be statistically analysed. Normally only a few analyses can be made, in which case five soil samples can be taken from each of three areas of a habitat. If the five samples from each area are bulked together and a sub-sample analysed, then three analyses from a given habitat will be available.

Little information is available to indicate the best time of the year for sampling. This time will depend on such factors as climate, vegetation and microbiological programme. In general the natural soil microflora can be expected to be most active towards the end of the growing season of the vegetation cover.

After sampling it is recommended that soils should be kept as cool as is practicable, but not lower than the normal refrigeration temperatures of about 4°C. The samples should not be bulked and sub-sampled until immediately prior to the analysis. During the period between the sampling and the analysis they should be kept in the sampling device or placed in plastic bags which are preferably impermeable to gases and moisture. If the core is wrapped in the plastic bag so that the top layer of soil remains open to the atmosphere as it was in the field, then changes in the concentration of gases in the sample will be reduced to a minimum.

At the time of analysing, the outside parts of samples can be removed with a knife leaving the less disturbed centre parts for the analysis. Prior to analysing, mineral soils are usually passed through a sterile 2 mm sieve and organic soils mechanically macerated in sterile water.

VI Conclusion

Three main general points emerge from this contribution on methods for studying soil microorganisms.

First, a great deal of work remains to be done on the development of techniques to allow accurate studies on microbial productivity in soil and litter.

Second, in the application of any of the techniques for 'standing-crop' assessments of soil microorganisms described above, slavish adherence to all the published details of any of these techniques should be avoided. Each technique must be modified, through thorough preliminary testing, to suit the particular soil under study.

Third, care must be taken to obtain sufficient ancillary information to allow microbiological data to be expressed in appropriate terms (e.g. per gm, dry wt., per unit volume, per metre^2, per unit of internal surface area of the soil), terms which allow microbiological data to be interpreted by other groups of soil biologists.

Acknowledgements

The authors wish to thank all the soil microbiologists who attended the UNESCO-IBP Symposium (Paris, 1967) for their active participation in discussions upon which this contribution is based.

References

The following references have been mentioned in the text or are useful reviews. It is suggested that they be consulted before deciding on the precise technique to be adopted. Space does not permit a more extensive list but the bibliographies at the end of each paper will serve as a further introduction to the literature.

ALEXANDER M. (1961). *Introduction to Soil Microbiology.* New York, Wiley.

ARISTOVSKAYA T.V. & PARINKINA O.M. (1962). Study of microbe patterns of soils of Leningrad oblast. *Mikrobiologiya,* **31,** 385.

BRIERLEY W.B., JEWSON S.T. & BRIERLEY M. (1928). The quantitative study of soil fungi. *1st Int. Cong. Soil Sci.* **3,** 48.

BROWN J.C. (1958). Fungal mycelium in dune soils by a modified impression slide technique. *Trans. Brit. mycol. Soc.* **41,** 81.

BURGES A. & NICHOLAS D.P. (1961). Use of soil sections in studying amount of fungal hyphae in soil. *Soil Sci.* **92,** 25.

EGDELL J.W., CUTHBERT W.A., SCARLETT C.A., THOMAS S.B. & WESTMACOTT M.H. (1960). Some studies of the colony count technique for soil bacteria. *J. appl. Bact.* **23,** 69.

EGGINS H.O.W. & PUGH G.J.F. (1962). Isolation of cellulose-decomposing fungi from soil. *Nature, Lond.* **193,** 94.

GABE D.R. (1961). Capillary method for studying microbe distribution in soils. *Soviet Soil Science,* **1,** 81.

GAMS W. & DOMSCH K.H. (1967). Beiträge zur Anwendung der Bodenwaschtechnik für die Isolierung von Bodenpilzen. *Arch. Mikrobiol.* **58,** 134.

GARRETT S.D. (1963). *Soil Fungi and Soil Fertility.* Macmillan Co., London.

GRAY T.R.G. & BELL T.F. (1963). The decomposition of chitin in soil. pp. 222–230. In: *Soil Organisms.* Doeksen and van der Drift (eds.). N. Holland Publishing Co.

HARLEY J.L. & WAID J.S. (1955). A method of studying active mycelia on living roots and other surfaces in soil. *Trans. Brit. mycol. Soc.* **38,** 104.

HERING T.F. (1966). An automatic soil washing apparatus for fungal isolation. *Plant and Soil,* **25,** 195.

HERING T.F. & NICHOLSON P.B. (1964). A clearing technique for the examination of fungi in plant tissue. *Nature,* **201,** 942.

HILL I.R. & GRAY T.R.G. (1966). Magnetic stirring as a method of dispersing soil bacteria in diluents. *Soil Biol.* **5,** 12.

JAMES N. (1958). Soil extract in soil microbiology. *Can. J. Microbiol.* **4,** 363.

JAMES N. & SUTHERLAND M.L. (1939). The accuracy of the plating method for estimating numbers of soil bacteria, actinomycetes and fungi in the dilution plated. *Can. J. Res. C.* **17**, 72.

JENSEN V. (1962). Studies on the microflora of Danish beech forest soils. I. The dilution plate count technique for enumeration of bacteria and fungi in soil. *Zentbl. Bakt. Parasit de., Abt. II*, **116**, 13.

JENSEN V. (1968). The plate count technique. In: *The Ecology of Soil Bacteria.* Gray & Parkinson (eds.). Liverpool University Press.

JOHNSON L.F. (1957). Effect of antibiotics on the numbers of bacteria and fungi isolated from soil by the dilution method. *Phytopathology,* **47**, 630.

JONES P.C.T. & MOLLISON J.E. (1948). A technique for the quantitative estimation of soil microorganisms. *J. gen. Microbiol.* **2**, 54.

MARTIN J.P. (1950). The use of acid, rose bengal and streptomycin in the plate method for estimating soil fungi. *Soil Sci.* **69**, 215.

MATHUR S.P. & PAUL E.A. (1967). Microbial utilization of soil humic acids. *Can. J. Microbiol.* **13**, 573.

MINDERMAN G. (1956). New techniques for counting and isolating free living nematodes from small soil samples and from oak forest litter. *Nematologica,* **1**, 216.

MONTÉGUT J. (1960). Value of the dilution method. pp. 43–99 in: *The Ecology of Soil Fungi.* Parkinson & Waid (eds.). Liverpool University Press.

NAGEL-DE BOOIS H.M. & JANSEN E. (1967). Hyphal activity in mull and mor of an oak forest. In: *Progress in Soil Biology.* Graff & Satchell.

NICHOLAS D.P. & PARKINSON D. (1967). A comparison of methods for assessing the amount of fungal mycelium in soil samples. *Pedobiologia,* **7**, 23.

ORCUTT F.S. (1940). Methods for more accurately comparing plate counts of soil bacteria. *J. Bact.* **39**, 100.

PARKINSON D. (1967). Soil microorganisms and plant roots. pp. 449–478 in: *Soil Biology.* Burges & Raw (eds.). Academic Press.

PERFIL'EV B.V. & GABE D.R. (1961). *Capillary Methods in Studying Microorganisms.* (In Russian.) Academy of Sciences, USSR.

POCHON J. (1954). *Manuel technique d'analyse microbiologique du sol.* Masson, Paris.

SKINNER F.E., JONES P.C.T. & MOLLISON J.E. (1952). A comparison of a direct and a plate-counting technique for the quantitative estimation of soil microorganisms. *J. gen. Microbiol.* **6**, 261.

SMITH N.R. & WORDEN S. (1925). Plate counts of soil microorganisms. *J. agric. Res.* **31**, 501

TRIBE H.T. (1957). Ecology of microorganisms in soils as observed during their development upon buried cellulose film. pp. 287–298 in: *Microbial Ecology, Seventh Symposium Soc. gen. Microbiol.* Williams & Spicer.

WAID J.S. & WOODMAN M.J. (1957). A method of estimating hyphal activity in soil. *Pedologie,* VII. no. spec., 155.

WARCUP J.H. (1950). The soil plate method for isolation of fungi from soil. *Nature, Lond.* **166**, 117.

WARCUP J.H. (1951). The ecology of soil fungi. *Trans. Brit. mycol. Soc.* **34**, 376.

WARCUP J.H. (1960). Methods for isolation and estimation of activity of fungi in soil. pp. 3–21 in: *The Ecology of Soil Fungi.* Parkinson & Waid (eds.). Liverpool University Press.

WARCUP J.H. (1967). Fungi in soil. pp. 51–110 in: *Soil Biology*. Burges & Raw (eds.). Academic Press.

WILLIAMS S.T., PARKINSON D. & BURGES N.A. (1965). An examination of the soil washing technique. *Plant and Soil,* 22, 167.

WILLIAMS S.T. & DAVIS F.L. (1965). Use of antibiotics for selective isolation and enumeration of actinomycetes in soil. *J. gen. Microbiol.* 38, 251.

WITKAMP M. (1960). Seasonal fluctuations of the fungus flora in mull and mor of an oak forest. *Meded.* 46 ITBON, Arnhem.

5

Protozoa

O.W. HEAL

The object of the present chapter is to indicate the most suitable methods available for the study of production and energy flow in protozoan populations. The main information required in such studies are estimates of:

(a) numbers and biomass

(b) birth and death rate

(c) food consumption and utilization

(d) respiration.

These four aspects have been examined to varying extents and although adequately tested methods are available for (a) and (d), there are no recognized methods for (b) and only a few potentially useful methods for (c). A wider review of the available and potential methods and the problems involved in these studies is given by Heal (1970). The present paper describes the most suitable methods currently available.

Unlike most other animals, protozoa are readily cultured. As a result there is much available information for individual species on their structure, physiology, life cycles and behaviour. Although relatively few soil species have been studied, the information and techniques show great potential for laboratory studies on population dynamics, behaviour in relation to environmental factors, qualitative and quantitative relationships with the microflora, and their metabolism. This information can be of great value in the understanding of the activities of protozoa *if* we can obtain adequate information on field populations. Lack of field techniques provides the biggest technical problem in protozoan studies.

Four distinct taxonomic groups of protozoa (ciliates, flagellates, naked amoebae and testate amoebae) are probably equally important in soil although in a particular soil one may be dominant. They have been studied to varying extents and recent reviews by Nikolyuk (1964, 1969), Pussard (1967), Noland

51

and Gojdics (1967) and Stout and Heal (1967) refer to many available data. Because of their different characteristics, the four groups often require particular methods of study.

1 Numbers

A range of techniques have been developed to estimate numbers of protozoa in soil and a number of more recent estimates are given in Appendix 1. The methods fall into three categories:
  (i) direct counting of soil
  (ii) extraction from soil
  (iii) culture from a soil dilution series.
Dilution methods are the best available for estimating numbers of protozoa despite the true statement of Bunt and Tchan (1955) that 'Direct microscopy is the logical method to overcome the selectivity of culture techniques'. Unfortunately although various direct methods have been tried, none have been rigorously tested, widely used or shown to be applicable to a wide range of protozoa. (Bunt and Tchan, 1955; Varga, 1959; Volz, 1951; Rosa, 1962; Schönborn, 1962; Heal, 1964a; Couteaux, 1967; Haarlov and Weis-Fogh, 1952–53; Minderman, 1956; Alexander and Jackson, 1955). Most of these methods are useful for Testacea with their resistant tests but preparation of the soil for examination usually makes other protozoa unrecognizable or unidentifiable. Extraction of active ciliates by electromigration (Hairston, 1964) or by temperature gradients (Uhlig, 1964) shows considerable potential but is not adequately developed. The extraction of Testacea by bubbling gas through soil (Decloitre, 1966) has recently been shown to be inefficient (Couteaux, 1967).

For the above reasons, dealt with more fully by Heal (1970), the dilution culture method originated by Cutler (1920) and developed by Singh (1946, 1955) is still the best available for estimating numbers of naked amoebae ciliates and flagellates, although modifications are necessary to suit particular soil or other requirements. This method is not adequate for Testacea but the direct methods of Jones and Mollison (Heal, 1964a) and Couteaux (1967) give good results for this group.

Singh dilution culture

The basic method, described by Singh (1946, 1955) is as follows: A 10 g sample of soil is shaken with 50 ml of normal salt solution giving a dilution of 1/5. From this, a series of twofold dilutions are made ranging from 1/10 to

1/18,920. Samples (0·05 ml) from these dilutions are inoculated on to petri dishes. Each petri dish contains eight glass rings embedded in NaCl agar. A thick suspension of an edible bacterium (*Aerobacter aerogenes*) is spread on the agar inside each ring. Protozoa in the soil suspension inoculated into each ring feed on the excess food and produce a large colony which can be readily recognized. Each ring is recorded as positive or negative after about 14 days and the number of protozoa per gram of soil is calculated from the total count of negative rings using statistical tables. This estimates active plus encysted protozoa. To estimate the numbers of encysted protozoa the soil is treated with 2% HCl for 24 hours to kill active forms, and after neutralizing the HCl the count is repeated.

This basic method has been modified by various workers for particular soils and organisms and the following comments are relevant.

1. Singh (1946) estimated the method to be 64–73% efficient for amoebae by adding known numbers to sterilized soil, then estimating them by the dilution culture method. This basic method, however, is probably less efficient for free-swimming protozoa because agar restricts their activity. Other workers have increased the available moisture by retaining agar and water in small cups (Stout, 1962) or repeatedly moistening the agar (Darbyshire and Greaves, 1967) but no assessment of the effectiveness of such modifications has been made.

2. With mineral soils protozoa are apparently adequately suspended in the soil dilutions by shaking (Singh, 1946; Darbyshire and Greaves, 1967) but in litter or organic soils protozoa may be trapped in large organic particles. The author found that estimates of numbers of amoebae in litter increased four-fold with light homogenizing which disintegrates the litter.

3. *Aerobacter aerogenes* supports growth of a wide range of protozoa and is probably the best available food (Singh, 1946). However, the provision of a known food can be omitted, Darbyshire and Greaves (1967) and Stout (1962) obtaining good results using soil extract agar or non-nutrient agar to allow the growth of the soil microflora as food for the protozoa. Bacteria must be added after treatment with HCl for cyst counts.

4. Sterilizable polypropylene rings are preferable to glass rings (Darbyshire and Greaves, 1967).

5. It is usually necessary to vary the method to suit the soil or conditions of the study. Wherever this is done tests should be made to find which modifications produce the highest reproducible counts. Samples are frequently bulked and the validity of this procedure should be tested (Singh, 1949).

Direct counts for Testacea

The resistant tests of Testacea allow them to withstand the preparations used for direct methods. This is particularly important as they rarely occur in dilution culture methods. The Jones and Mollison (1948) soil+agar suspension films used for bacterial and fungal counts (Parkinson *et al.* in this handbook) give high counts of Testacea (Heal, 1964a; Volz, 1964). Slight modifications to the basic techniques may be necessary, e.g. increasing the concentration of soil to increase numbers of Testacea examined. An alternative method which has produced high counts and is apparently simpler than the Jones and Mollison method is described by Couteaux (1967). In this a soil sample is fixed (Bouin-Hollande), stained (Ponceau de xylidine), diluted with distilled water then a small sub-sample is filtered through a 25 mm diameter Millipore filter. The filter is then cleared, mounted and Testacea counted microscopically. Mm. Couteaux (pers. comm.) has found that 1–2 mg of soil per filter provides maximum counts.

2   Biomass and calorific determination

The weight of a field population of protozoa can rarely be measured directly and an estimate is usually obtained from the weight per individual × the number present. Two main methods are available for estimating the weight of an individual and a number of published estimates of the weights of various species are given in Appendix 2.

(i) For most studies estimates of the wet weight of an individual can be made by calculating cell volume and assuming a specific gravity of 1·0. Such estimates can be made on individuals isolated from the field or from laboratory cultures.

Cell volume is often calculated from linear measurements assuming a simple geometrical shape. An excellent example is given by Hull (1961) for *Podophrya collini* and *Tetrahymena pyriformis*. A compression chamber is frequently used to immobilize and flatten the cell before measurement. Cameron and Prescott (1961) constructed a simple chamber with strands of glass wool, 8 μ in diameter, attached to a microscope slide. A small drop of culture containing the specimen (*Tetrahymena pyriformis*) was placed on the slide and flattened with a cover slip. The compressed cell was photographed, the area measured with a planimeter and used to calculate the volume. Reuter (1963) obtained greater accuracy when ciliates were forced into a thin tapering capillary and the volume calculated from their length and the internal diameter of the capillary assuming a cylindrical shape.

(ii) Measurement of the dry weight per individual can be obtained by cropping, drying and weighing cultures which have been counted. This method is widely used in laboratory studies but a check must be made that the organism in culture is the same size as that in the field because protozoa vary considerably in size depending on the age of the culture and the conditions (Adolph, 1929; Ormsbee, 1942; Kimball *et al.*, 1959).

The mass cultures used for dry weight estimation can provide adequate material for calorific determination by standard procedures.

3 Annual production

Although the numbers of protozoa may be similar on consecutive sampling dates, a large number of individuals may have been produced and died between the samplings. With potential doubling times of less than one day, the population turnover of protozoa is much greater than in most animal groups.

Intensive sampling is not necessarily the best method of estimating production and most counting techniques are very time consuming and prevent sampling at more than weekly intervals. It is better to sample less intensively but to obtain, at the same time, an estimate of the generation time or death rate of the organisms. This provides more information on the biology of the organisms as well as allowing production to be estimated. If the generation time in soil is known a formula such as that of Calkorskaja reported in Kajak (1967) can be used. In this

$$P = \frac{N0 + N1}{2} \times \frac{1}{D} \times t$$

where P = production, $N0$ and $N1$ are initial and final numbers (or biomass) in time t and D is the period of doubling of an individual.

Unfortunately there are no obvious techniques to provide accurate estimates of generation time or death rate in soil and the following comments are entirely speculative. Direct observation of live individuals for long periods, possibly with time-lapse cine photography, could provide useful information. To allow observation it may be necessary to introduce a transparent surface such as a cellulose film into the soil and make recordings after colonization by soil microorganisms. Measurement of the rate of disappearance of marked individuals from a population is also possible if a suitable vital marker can be found. Where it is possible to recognize individual protozoa in the process of division it should be possible to estimate production if (a) the percentage of the population dividing at any moment can be measured and (b) the length of the time taken by the division process is known.

4   Energy flow

An estimate of the amount of energy used by protozoa can show the size of their contribution to the total energy flow in a particular ecosystem. To obtain this it is necessary to know the field population, its production and the energy budget of individual animals. The energy budget is best expressed as follows, the components being given in calories.

$$\text{Amount of food} = \underbrace{\text{Amount of protozoa} + \text{Amount lost in}}_{\text{Amount assimilated (A)}} + \text{Amount}$$

Amount of food = Amount of protozoa + Amount lost in + Amount
  consumed (C)   produced (P)   respiration (R)   egested (F)

Amount assimilated (A)

For protozoa, C, R and P can be measured directly, F is usually obtained by calculation because collection of protozoan faeces is impracticable. This energy budget can be obtained readily from laboratory populations with the associated dangers of later extrapolation, to the field. Only the main principles of available methods are given here because details will differ considerably depending on the organisms involved and facilities available.

Measurement of consumption and production

Two main methods of studying the energy budget of protozoa are available.
  (i) Direct: numbers of food organisms ingested and protozoa produced in a culture are counted directly and converted to dry weight or calories (Nero *et al.*, 1964; Heal, 1967; Salt, 1967; Curds, 1968).
 (ii) Indirect: for example using radiotracers. Labelled food is added to the culture and the increase in radioactivity of the protozoan cells is used to estimate food consumption (Lee *et al.*, 1966).

(i) *Direct methods*

From counts of the numbers of food organisms and protozoa at the beginning and end of an experimental culture the food consumption and protozoa production can be estimated if (a) there is no reproduction or non-predatory death of the food organism and (b) no death of protozoa. Significant reproduction of food organisms can be prevented by using organisms with slow reproductive rates and non-nutrient media or by incorporating a non-toxic substance, e.g. antibiotics which prevents reproduction of the food but does not affect the protozoa.

A correction factor can be applied if the reproductive rate of the food is known, usually from control cultures. However, because of the absence of protozoa, and therefore their excretory products, the nutrient status of test and control cultures is different. This also applies when estimating, from control cultures, non-predatory death of the food and death of protozoa, but these methods are the best available (Salt, 1967). With certain organisms, especially large ones, non-predatory death can be recognized by direct observation. It is an advantage for experiments to be carried out over the shortest possible period, to reduce these complicating factors.

The quantities of food and protozoa can be estimated by various methods; direct counting with counting chambers, e.g. haemocytometers, electronic particle counters, measurement of tubidity, dry weight or total nitrogen. The biggest difficulty is to distinguish between the three types of particle present in the culture; food organisms, protozoa and faeces. Direct observation in counting chambers, although tedious, is simple and allows observation of the state of the organisms and the separation of food and faeces. Particles of different size can be separated by an electronic counter or by differential centrifugation but separation of food and faeces may be difficult.

Two other methods are suitable for particular protozoa. (a) Salt (1967) used time-lapse cine photography to record, on a single frame, the number of large ciliate predators (*Woodruffia*) and prey (*Paramecium*) in a micro-culture. (b) Numbers of bacteria consumed by ciliates can be estimated by counting the number of bacteria per food vacuole and recording the number of vacuoles formed per unit time (Harding, 1937; Bahr, 1954).

(ii) *Indirect methods*

The basic principles by which radiotracers can be used to measure ingestion and assimilation are well known. Although well adapted for use with protozoa, such methods have rarely been used in ecological studies (Lee *et al.*, 1966; Coleman, 1964).

After trying various methods, Lee *et al.* (1966) used radiotracers to measure food consumption in marine Foraminifera in a range of conditions. Food organisms (algae, yeasts, bacteria) were grown in labelled 32P or 14C and known numbers added to the Foraminifera cultures. At various intervals between 4 and 32 days the Foraminifera were cropped, counted and their radioactivity assessed. From this measure the number of organisms consumed per animal was calculated. Although no estimates were made of the dry weight or calorific value of food or protozoa, such measures can be readily obtained. However, the proportion of the radiotracer retained by the predator depends, among other things, on the assimilation rate which varies with type and concentration of food and with time. Despite this the method has potential for measuring food utilization.

Measurement of respiration

Hundreds of measurements of respiration rates of protozoa are published, all from laboratory cultures. Some examples from various living protozoa, not necessarily soil forms, are given in Appendix 3. The methods which are well tried and can easily be applied to soil protozoa fall into two categories which will not be described in detail in protozoa.

(i) *Micro-respirometers.* The use of Cartesian diver respirometers (Holter, 1943; Zajicek and Zeuthen, 1956) and similar flotation methods (Scholander *et al.*, 1952) allow the measurement of respiration of single cells or small numbers of individuals with a sensitivity as low as $0 \cdot 1$ μμl O_2/hr. These methods allow measurements to be made at different stages in the life cycle (Pigon, 1959; Stewart, 1964).

(ii) *Macro-respirometers.* Respiration of large cultures is readily measured by standard Warburg techniques (e.g. Ormsbee, 1942; Reich, 1948; Neff *et al.*, 1958) or with more recently developed oxygen diffusion probes (e.g. Cook, 1956).

Axenic cultures allow measurement of endogenous and exogenous respiration without the complication of the respiration of food organisms. However, it is not known whether respiration of axenic protozoa is the same as that of protozoa fed on live food.

Food consumption, protozoan production and respiration can be measured together if the culture is in a respirometer. Respiration of the food can be separated from that of the protozoa by calculation, using an independent estimate of the respiration of the food organism, in control cultures.

Discussion

In attempting to estimate the role of protozoa in soil, whether in terms of energy or a particular nutrient passing through the population it is necessary to know the numbers of protozoa present, their rates of production and death, their food and the way in which this is used. Many species are easily cultured and valuable information can be obtained concerning feeding and respiration, the latter particularly in axenic cultures. In cultures many factors such as cell age, concentration and medium affect respiration rates greatly (Appendix 3) and different strains of species may have different rates (Seaman, 1955). Thus although it is relatively easy to obtain measures of respiration, extrapolation to field conditions is dangerous. However, errors from laboratory studies on feeding and respiration may be small compared with those field estimates of numbers and production. Although there is a natural tendency to develop laboratory studies, because they are practicable and produce apparently exact results, greater effort should be made with field studies. Where manpower is limited a more accurate estimate of the energy budget of protozoa will be obtained by examining field populations in detail and using published data on food consumption and respiration, than by dividing the effort between field and laboratory studies. This is probably true also for other microfauna and microflora.

Estimates of annual production are far more important in microorganisms than for higher organisms. For vertebrates and large invertebrates relative production (annual production/standing crop) is probably about 0·5—1·0, for small invertebrates about 1·0—4·0, but for microorganisms about 10–100. Unfortunately at present we have not attempted production estimates for those organisms in which they are most important. In most faunal studies a figure for the annual energy flow through a population (population metabolism) has been obtained using the sum of the monthly population estimate × respiration per individual, with a correction for temperature. This may be the best available figure in many IBP studies. It will be reasonably accurate for many invertebrates as about 90% of assimilated energy is used in respiration (Macfadyen, 1963). However, protozoa probably convert a larger percentage of assimilated food to protoplasm (Heal, 1967; Coleman, 1964; Curds, 1968). They are permanently in a growth phase, division occurring when a certain cell size is reached, therefore they lack an adult stage in which growth is minimal and resemble the young stages of metazoa in which growth efficiency is high (Engelmann, 1961).

Although estimates can be obtained of the amount of food consumed by particulate feeding protozoa, non-particulate food may be important in certain habitats, particularly those rich in organic matter. Many flagellates are capable of osmotrophy and amoebae and ciliates, by means of pinocytosis, ingest organic molecules in liquid. The extent of these functions in soil is completely unknown and will probably remain so until delicate tracer studies can be made.

Summary

1. Soil protozoa belong to four main groups (flagellates, ciliates, naked amoebae and testate amoebae) which often require very different methods of study.

2. It is preferable to estimate numbers by direct observation of the soil or extraction but the dilution culture method of Singh, with modifications to improve estimates of free-swimming forms, is probably the best available technique.

3. Testacea are the only group for which good direct counting methods are available, the methods of Jones and Mollison, and Couteaux being most suitable.

4. With short generation times, turnover in protozoan populations is high, but no methods are available to estimate reproduction or death rates in the field.

5. Well developed culture techniques allow accurate measurements of ingestion, assimilation, production and respiration to be made in the laboratory.

6. Excellent opportunities exist for accurate laboratory studies on protozoa but because field data are so inaccurate, emphasis should be placed on the latter wherever possible.

Acknowledgements

I am very grateful to J.F. Darbyshire (U.K.), M. Ertl (Czechoslovakia), J.E. Satchell (U.K.) and J.D. Stout (New Zealand) for constructive criticism of the text.

References

ADOLPH E.F. (1929). The regulation of adult body size in the Protozoan *Colpoda. J. exp. Zool.* **53,** 269–311

ALEXANDER F.E.S. & JACKSON R.M. (1955). Preparation of sections for study of soil microorganisms. In: *Soil Zoology.* D.K.McE. Kevan (ed.). Butterworths, London. 433–441.

ATLAVINYTE D., EITEMINAVICIUTE I., GRIGELIS A., LIEPINIS A., STRAZDIENE V. & SLEPETIENE J. (1967). Biocoenosis dynamics of invertebrate fauna in eroded soils under crops. In: *Progress in Soil Biology*. O. Graff & J.E. Satchell (eds.). North Holland Publ. Co., Amsterdam. 353–359.

BAHR H. (1954). Untersuchungen uber die Rolle der Ciliaten als Bakterienvernichter im Rahmen der biologischen Reinigung des Abwassers. *Z. Hyg. Infekt Krankh.* **139**, 160–181.

BAND R.N. (1959). Nutritional and related biological studies on the free-living soil amoeba, *Hartmannella rhysodes. J. gen. Microbiol.* **21**, 80–95.

BRZEZINSKA-DUDZIAK B. (1954). Oznaczanie ilości ameb w glebie wg zmo dyfikowanej metody Singha. *Acta microbiol. pol.* **3**, 121–124.

BUNT J.S. & TCHAN Y.T. (1955). Estimation of protozoan populations in soils by direct microscopy. *Proc. Linn. Soc. N.S.W.* **80**, 148–153.

CALKINS G.N. & SUMMERS F.M. (1941). *Protozoa in Biological Research*. Columbia Univ. Press, New York.

CAMERON I.L. & PRESCOTT D.M. (1961). Relations between cell growth and cell division v. cell and macronuclear volumes of *Tetrahymena pyriformis* H.S.M. during the cell life cycle. *Expl. Cell Res.* **23**, 361–372.

CHAPMAN-ANDRESEN C. & DICK D.A.T. (1961). Volume changes in the amoeba *Chaos chaos* (L.). *C. r. Trav. Lab. Carlsberg*, **32**, 265–289.

COLEMAN G.S. (1964). The metabolism of *Escherichia coli* and other bacteria by *Entodinium caudatum. J. gen. Microbiol.* **37**, 209–223.

CONNER R.L. & CLINE S.G. (1967). Some factors governing respiration, glucose metabolism and iodoacetate sensitivity in *Tetrahymena pyriformis. J. Protozool.* **14**, 22–26.

COOK J.R. (1966). Adaptations to temperature in two closely related strains of *Euglena gracilis. Biol. Bull.* **131**, 83–93.

COUTEAUX M.M. (1967). Une technique d'observation des Thecamoebiens du sol pour l'estimation de leur densité absolue. *Rev. Ecol. Biol. Sol*, **4**, 593–596.

COUTEAUX M.M. (1969). Etude de la communauté de Thécamoebiens d'une chênaie à luzule (Moyenne-Belgique). *C.R. Acad. Sc. Paris*, **269**, 335–338.

CURDS C.R. & COCKBURN A. (1968). Studies on the growth and feeding of *Tetrahymena pyriformis* in axenic and monoxenic culture. *J. gen. Microbiol.* **54**, 343–358.

CUTLER D.W. (1920). A method for estimating the number of active protozoa in the soil. *J. agric. Sci. Camb.* **10**, 135–143.

CUTLER D.W. & CRUMP L.M. (1927). The qualitative and quantitative effects of food on the growth of a soil amoeba (*Hartmannella hyalina*). *Br. J. exp. Biol.* **5**, 155–165.

CUTLER D.W., CRUMP L.M. & SANDON H. (1922). A quantitative investigation of the bacterial and protozoan population of the soil with an account of the protozoan fauna. *Phil. Trans. R. Soc. B.* **211**, 317–350.

DANFORTH W.F. (1967). Respiratory metabolism. In: *Research in Protozoology*. T.T. Chen (ed.). Pergamon Press, Oxford. **1**, 201–306.

DARBYSHIRE J.F. & GREAVES M.P. (1967). Protozoa and bacteria in the rhizosphere, of *Sinapis alba* L., *Trifolium repens* L. and *Lolium perenne* L. *Canad. J. Microbiol.* **13**, 1057–1068.

DECLOITRE L. (1966). Comment compter le nombre de Thecamoebiens dans une recolte *Limnologica (Berlin)*. **4**, 489–492.

ENGELMANN M.D. (1961). The role of soil arthropods in the energetics of an old field community. *Ecol. Monogr.* **31**, 221–238.

FORMISANO M. (1957) Ricerche microbiolgiche sulla 'rizostera' delle piante coltivate rei terreni della compania. *Annali Fac. Agr. Portici*, **22**, 1–34.

GAW H.Z. (1941). Soil protozoa in some Chinese soils. *Nature, Lond.* **147**, 390.

HAARLOV N. & WEIS-FOGH T. (1952–53). A microscopical technique for studying the undisturbed texture of soils. *Oikos*, **4**, 44–57.

HAIRSTON N.G. (1964). *Paramecium* ecology: electromigration for field samples and observations on density. *Ecology*, **45**, 373–376.

HARDING J.P. (1937). Quantitative studies on the ciliate *Glaucoma*. I. The regulation of the size and the fission rate by the bacterial food supply. *J. exp. Biol.* **14**, 422–430.

HEAL O.W. (1964a). The use of cultures for studying Testacea (Protozoa: Rhizopoda) in soil. *Pedobiologia*, **4**, 1–7.

HEAL O.W. (1964b). Observations on the seasonal and spatial distribution of Testacea (Protozoa: Rhizopoda) in *Sphagnum*. *J. Anim. Ecol.* **33**, 395–412.

HEAL O.W. (1965). Observations on testate amoebae (Protozoa: Rhizopoda) from Signy Island, South Orkney Islands. *Br. Antarctic Surv. Bull.* No. 6, 43–47.

HEAL O.W. (1967). Quantitative studies on soil amoebae. In: *Progress in Soil Biology.* O. Graff & J.E. Satchell (eds.). North Holland Publ. Co., Amsterdam. 120–125.

HEAL O.W. (1970). Methods of study of soil protozoa. In: *Methods of Study in Soil Ecology.* J. Phillipson (ed.). UNESCO, 119–126.

HOLTER H. (1943). Technique of the Cartesian Diver. *C. r. Trav. Lab. Carlsberg, ser. chim.* **24**, 399–478.

HULL R.W. (1961). Studies on Suctorian protozoa: the mechanism of ingestion of prey cytoplasm. *J. Protozool.* **8**, 351–359.

HUNTER F.R. & LEE J.W. On the metabolism of *Astasia longa* (John). *J. Protozool.* **9**, 74–78.

HUTCHENS J.O. (1941). The effect of the age of the culture on the rate of oxygen consumption and the respiratory quotient of *Chilomonas paramecium*. *J. cell comp. Physiol.* **17**, 321–332.

JONES P.C.T. & MOLLISON T. (1948). A technique for the quantitative estimation of soil microorganisms. *J. gen. Microbiol.* **2**, 54–69.

KAJAK Z. (1967). Remarks on methods of investigating benthos production. *Ekol. pol. B.* **13**, 173–195.

KAWAKAMI T. (1964). Factors influencing excystation in *Tillina magna*. *J. Protozool. Suppl.* **11**, 10.

KIDDER G.W. & DEWEY V.C. (1951). The biochemistry of ciliates in pure culture. In: *Biochemistry and Physiology of Protozoa.* A. Lwoff (ed.). Academic Press, New York. **1**, 323–400.

KIMBALL R.F., CASPERSSON T.O., SVENSSON G. & CARLSON L. (1959). Quantitative cytochemical studies on *Paramecium aurelia*. I. Growth in total dry weight measured by the scanning interference microscope and X-ray absorption methods. *Expl. Cell Res.* **17**, 160–172.

KOFFMANN M. (1932). Die Mikrofauna des Bodens und ihr verhältnis zu anderen Bodenimikroorganismen. Die Rolle der Bodenprotisten bei den Mikrobiologischen Prozessen im Boden. *Proc. 2nd Int. Cong. Soil Sci. Leningrad-Moscow*, **3**, 268–271.

LEE J.L., MCENERY M., PIERCE E., FREUDENTHAL H.D. & MULLER W.A. (1966). Tracer experiments in feeding littorial Foraminifera. *J. Protozool.* **13**, 659–670.

MACFADYEN A. (1963). Heterotrophic productivity in the detritus food chain in soil. *Proc. XVI Int. Cong. Zool., Washington.* **4**, 318–329.

MINDERMAN G. (1956). The preparation of microtome sections of unaltered soil for the study of soil organisms in situ. *Pl. Soil*, **8**, 42–48.

NEFF R.J., NEFF R.H. & TAYLOR R.H. (1958). The nutrition and metabolism of a soil amoebae *Acanthamoeba* sp. *Physiol. Zoöl.* **31**, 73–91.

NERO L.C., TARVER MAE G. & HEDRICK L.R. (1964). Growth of *Acanthamoeba castellani* with the yeast *Torulopsis famata*. *J. Bact.* **87**, 220–225.

NIKOLJUK V.F. (1964). [Soil dwelling Protozoa and their biological importance]. *Pedobiologia*, **3**, 259–273.

NIKOLJUK V.F. (1969). Some aspects of the study of soil protozoa. *Acta protozool.* **7**, 99–109.

NOLAND L.E. & GOJDICS M. (1967). Ecology of free-living protozoa. In: *Research in Protozoology*. T.T. Chen (ed.). Pergamon Press, Oxford. **2**, 216–266.

ORMSBEE R.A. (1942). The normal growth and respiration of *Tetrahymena geleii. Biol. Bull.* **82**, 423-437.

PACE D.M. (1946). Investigation on respiration in the colourless flagellate *Chilomonaₒ paramecium* with special emphasis on the vitamins involved in the respiratory mechanism. *Yb. Am. Phil. Soc.* 1946. 160–162.

PARKINSON D., GRAY T.R.G., HOLDING J. & NAGEL-DE-BOOIS H.M. (1971) Methods for the quantitative study of heterotrophic soil micro-organisms. In: *Methods of Study in Quantitative Soil Ecology*. J. Phillipson (ed.). I.B.P. Handbook No. 18. Blackwells, Oxford.

PIGON A. (1959). Respiration of *Colpoda cucullus* during active life and encystment. *J. Protozool.* **6**, 303–308.

PRESCOT D.M. (1955). Relations between cell growth and cell division. 1. Reduced weight, cell volume, protein content and nuclear volume of *Amoeba proteus* from division to division. *Expl. Cell Res.* **9**, 328–337.

PUSSARD M. (1967). Les protozoaires du sol. *Ann. Epiphyties,* **18**, 335–360.

REICH K. (1948). Studies on the respiration of an amoeba, *Mayorella palastinensis. Physiol. Zoöl.* **21**, 390–412.

REUTER J. (1963). The internal concentration in some Hypotrichous ciliates and its dependence on the external concentration. *Acta zool. fenn.* **104**, 1–94.

REYNOLDS H. & WRAGG JUNE B. (1962). Effect of type of carbohydrate on growth and protein synthesis by *Tetrahymena pyriformis*. *J. Protozool.* **9**, 214–222.

ROSA K. (1957). Vyzkum microcdafonu ve smrkovem porostu na Pradedu. *Sb. pvir. Spol. Mor. Ostrave.* **18**, 17–75.

ROSA K. (1962). Mikroedafon im degradierten Kieferbestand und im Topfen auf Tertiarem Sand im Nova Ves nei Ceske Budejovice. *Acta Univ. Carol. Biol. Suppl.* 1962. 7–30.

SALT G.W. (1967). Predation in an experimental protozoan population (*Woodruffia-Paramecium*). *Ecol. Monogr.* **37**, 113–144.

SEAMAN G.R. (1955). Metabolism of free-living ciliates. In: *Biochemistry and Physiology of Protozoa*. S.H. Hutner and A. Lwoff (eds.). Academic Press, New York. **2**, 91–158.

SCHONBORN W. (1962). Zur Okologie der sphagnikolen, bryokolen und terrikolen Testacean. *Limnologica (Berlın)*, **1**, 231–254.

SCHOLANDER P.F., CLAFF C.L. & SVEINSSON S.L. (1952). Respiratory studies of single cells. I. Methods. *Biol. Bull.* **102**, 157–177.

SINGH B.N. (1946). A method of estimating the number of soil Protozoa, especially amoebae, based on their differential feeding on bacteria. *Ann. appl. Biol.* **33**, 112–119.

SINGH B.N. (1949). The effect of artificial fertilizers and dung on the numbers of Amoebae in Rothamsted soils. *J. gen. Microbiol.* **3**, 204–210.

SINGH B.N. (1955). Culturing soil Protozoa and estimating their numbers in soil. In: *Soil Zoology*. D.K. McE.Kevan (ed.). Butterworths, London. 403–411.

SINGH B.N. & CRUMP L.M. (1953). The effect of partial sterilization by steam and formalin on the numbers of Amoebae in field soil. *J. gen. Microbiol.* **8**, 421–426.

STEWART J.M. (1964). The measurement of oxygen consumption in paramecia of different ages. *J. Protozool. Suppl.* **11**, 39–40.

STOUT J.D. (1962). An estimation of microfaunal populations in soils, and forest litter. *J. Soil Sci.* **13**, 314–320.

STOUT J.D. & HEAL O.W. (1967). Protozoa. In: *Soil Biology*. A. Burges and F. Raw (eds.). Academic Press, London. 149–195.

THIMANN K.V. & COMMONER B. (1940). A differential volumeter for micro-respiration measurements. *J. gen. Physiol.* **23**, 333–341.

UHLIG G. (1964). Eine einfache Methode zur Extraktion der vagilen mesopsammelen Mikrofauna. *Helgölander wiss. Merresunters* **11**, 178–185.

VARGA L. (1956). Adatok az Alföldi fasitott szikestalajok mikrofaunajanak ismeretehez. *Magy. tudom. Akad. Biol. agrartud. Osztál Közl.* **9**, 57–69.

VARGA L. (1959).Untersuchungen uber die Mikrofauna der Waldstreu einiger Waldtypen im Bukkgebugt (Ungarn). *Acta zool. hung.* **4**, 443–478.

VOLZ P. (1951). Untersuchungen uber die Microfauna des Waldbodens. *Zool. Jb. Abt. Sy.t. Oekol.* **79**, 514–566.

VOLZ P. (1964). Uber die soziologische und die physiognomische Forschungsnichtung in der Bodenzoologie. *Verh. Deutsch Zool. Ges. Kiel.* 522–532.

VON DACH H. (1942). Respiration of a colourless flagellate, *Astasia klebsii*. *Biol. Bull.* **82**, 356–371.

ZAJICEK J. & ZEUTHEN E. (1956). Quantitative determination of Cholinesterase activity in individual cells. *Expl. Cell Res.* **11**, 568–579.

ZEUTHEN E. (1943). A Cartesian diver micro-respirometer, with a gas volume of 0·1 μl. *C. r. Trav. Lab. Carlsberg ser chim.* **24**, 479–518.

ZEUTHEN E. (1953). Oxygen uptake as related to body size in organisms. *Q. Rev. Biol.* **28**, 1–12.

APPENDIX

APPENDIX 1. Some published estimates of numbers of protozoa in various soils.*

$\times 10^3$/g wet wt.	$\times 10^3$/g dry wt.	$\times 10^6$/m²	Method	Soil	Author
Total Protozoa					
403			Dilution	Fertilized irrigated	Nikoljuk (1964)
7·2			Dilution	N-fertilizer	Nikoljuk (1964)
3·7			Dilution	Unfertilized	Nikoljuk (1964)
4–27			Direct	Acid, brown earth under spruce	Rosa (1957)
50–100			Direct	Woodlands, rendzina soils	Varga (1959)
100			Direct	Various	Varga (1956)
0·06			Dilution	Sand } Range of soils,	Gaw (1941)
112·5			Dilution	Rice field clay } flagellates dominant	
Naked amoebae					
	2–51		Dilution	Cultivated sandy loam	Darbyshire and Greaves (1967)
	12–180		Dilution	Cultivated sandy loam, rhizosphere	Darbyshire and Greaves (1967)
1–54			Dilution	Sandy forest soil, unfertilized	Singh and Crump (1953)
	2–22		Dilution	Clay loam, unfertilized	Singh (1949)
	12–93		Dilution	Clay loam, mineral fertilizers	Singh (1949)
	15–146		Dilution	Clay loam, farmyard manure	Singh (1949)
12–231			Dilution	Clay loam	Cutler et al. (1922)
	421–1690		Dilution	Various	Brzezinska-Dudziak (1954)
	64		Dilution	Flooded rice field	Brzezinska-Dudziak (1954)
1–30			Dilution	Various	Formisano (1957)
1			Dilution	Podzolized sandy loam	Atlavinyte et al. (1967)
50			Direct	Various	Koffmann (1932)
Testate amoebae†					
		13–19	Direct	Woodland, brown earth, mull	Volz (1964)
		484–491	Direct	Woodland, brown earth, moder	Volz (1964)
		5–25	Direct	Woodland	Volz (1964)
17·7	35		Direct	Raw humus under oak and Vaccinium	Couteaux (1967)

APPENDIX 1. *Continued.*

	5–25	Direct	Beech and pine woodlands on sand	Schönborn (1962)
9–69	450–900	Direct	Grassland, mineral	Heal (1965)
2–73	190–1050	Direct	Woodland, brown earth, mull and moder	Heal (1964a, 1965)
0·3–8·1		Direct	Pinewood on sand	Rosa (1962)
Flagellates				
	0–30	Dilution	Cultivated sandy loam	Darbyshire and Greaves (1967)
	0–56	Dilution	Cultivated sandy loam, rhizosphere	Darbyshire and Greaves (1967)
70		Dilution	Cultivated	Singh (1946)
166–434		Dilution	Cultivated clay loam, monthly means	Cutler *et al.* (1922)
<90		Dilution	Podzolized sandy loam	Atlavinyte *et al.* (1967)
2–34		Dilution	Various	Formisano (1957)
100		Direct	Various	Koffmann (1932)
0·1–3·3		Direct	Pinewood on sand	Rosa (1962)
Ciliates				
	<0·28	Dilution	Cultivated sandy loam	Darbyshire and Greaves (1967)
	<0·75	Dilution	Cultivated sandy loam, rhizosphere	Darbyshire and Greaves (1967)
0·014–0·18		Dilution	Silt and clay loams	Stout (1962)
0·004		Dilution	Peaty sand	Stout (1962)
0·23 –1·45		Dilution	Beech litter	Stout (1962)
0·37		Dilution	Cultivated clay loam	Singh (1946)
1·0		Dilution	Podzolized sandy loam	Atlavinyte *et al.* (1967)
1·0 –23·0		Dilution	Various	Formisano (1957)
1·0		Direct	Various	Koffmann (1932)
0·04 –3·6		Direct	Pinewood on sand	Rosa (1962)

* Counts represent active and encysted protozoa. The proportion of encysted individuals varies greatly but is often about 50%.

† Live individuals only (=full tests).

See also more recent papers by Courteaux (1969) and Nikoljuk (1969).

APPENDIX 2. Some published estimates of the dry weight or volume ($\equiv$ wet weight) of various free-living protozoa.

Species	mg dry wt./10^6	mm³/10^6	Comments	Author
Rhizopods				
Chaos chaos		72000		Zeuthen (1953)
C. chaos		5500–15000		Scholander *et al.* (1952)
C. chaos		22000		Chapman-Andresen and Dick (1961)
Amoeba proteus		497	newly divided	Prescott (1955)
Mayorella palastinensis	1·16			Reich (1948)
Acanthamoeba sp.	0·37			Heal (1967)
Acanthamoeba sp.	0·84			Neff *et al.* (1958)
Hartmanella rhysodes	0·75		assuming 10% N content	Band (1959)
H. hyalina		83–288		Cutler and Crump (1927)
Difflugia sp.		833		Zeuthen (1943)
Cryptodifflugia vulgaris		3	9 species given	Volz (1951)
Arcella discoides		150	within this range	
Amphitrema flavum		11		Heal (1964b)
Hyalosphenia papilio		75		Heal (1964b)
Nebela tincta		97		Heal (1964b)
Flagellates				
Astasia kelbsii		5		*Scholander *et al.* (1952)
Euglena gracilis	0·8–2·2			Cook (1966)
Ciliates				
Tetrahymena pyriformis	15			*Danforth (1967)
T. geleii	5·3–9·0			Ormsbee (1942)
T. geleii		15–24		*Scholander *et al.* (1952)
T. pyriformis		16–33		Reynolds and Wragg (1962)

APPENDIX 2. *Continued.*

Ciliates—continued

				Source*
T. pyriformis		31		Hull (1961)
Paramecium caudatum		108–1000		Scholander *et al.* (1952)
P. aurelia	7·5–54			Kimball *et al.* (1959)
Bresslaua insidiatrix		15–35		Scholander *et al.* (1952)
Colpoda sp.		1500	16·5°C	Adolph (1929)
Colpoda sp.		1100	26·5°C	Adolph (1929)
Paraholosticha herbicola		118	trophic	Reuter (1963)
P. herbicola		69	cyst	Reuter (1963)
Oxytricha fallax		91	trophic	Reuter (1963)
O. fallax		43	cyst	Reuter (1963)
Euplotes taylori		56	trophic	Reuter (1963)
E. taylori		64	cyst	Reuter (1963)
Steinia inquieta		55	trophic	Reuter (1963)
S. inquieta		24	cyst	Reuter (1963)

* Source, where original papers not consulted.

APPENDIX 3. Some published estimates of the O₂ consumption of various free-living protozoa. Temperatures unless otherwise stated are between 20° and 28°C.

Species	$mm^3 O_2/hr/10^6$	$mm^3 O_2/hr/mg$ dry wt.	Comments	Author
Rizopods				
Chaos chaos	2000–6000			Scholander et al. (1952)
C. chaos	15000–56000			Zeuthen (1953)
Amoeba proteus	300–1400			*Scholander et al. (1952)
Mayorella palastinensis	11·7	10·1	trophic	Reich (1948)
M. palastinensis	2·2		cysts	Reich (1948)
Acanthamoeba sp.	10·1†			Neff et al. (1958)
Difflugia sp.	1160			Zeuthen (1943)
Flagellates				
Astasia klebsii	1·9–10		inorganic medium	von Dach (1942)
A. klebsii	<69·5		organic medium	von Dach (1942)
Astasia sp.	2400			*Calkins and Summers (1941)
A. longa		6·0–14·5	endogenous, assuming 10% N content	Hunter and Lee (1962)
Euglena gracilis	15–20		endogenous	Cook (1966)
E, gracilis	35–70		exogenous	Cook (1966)
E. gracilis	20–54		varies with centrifugation	*Danforth (1967)
Khawkinea kalli	2050			*Calkins and Summers (1941)
Chilomonas paramecium	663		exogenous	Pace (1946)
C. paramecium	325		endogenous	Pace (1946)
C. paramecium	17–26			*Calkins & Summers (1941)
C. paramecium	17–40		variation with age	Hutchens (1941)

APPENDIX 3. *Continued.*

Ciliates

Tetrahymena pyriformis		10	150 hrs old culture		*Kidder and Dewey (1951)
T. pyriformis		19·6	70 hrs old culture		*Kidder and Dewey (1951)
T. pyriformis		13·9	48 hrs old culture		*Kidder and Dewey (1951)
T. pyriformis	57–95				Conner and Cline (1967)
T. pyriformis		10–40	endogenous		*Danforth (1967)
T. pyriformis	50–271	6–35	various authors		*Ormsbee (1942)
T. pyriformis	433	81	exponential ⎫ with peptone		Ormsbee (1942)
T. pyriformis	633	79	stationary ⎭ with peptone		
T. pyriformis	140	26	exponential ⎫ without peptone		Ormsbee (1942)
T. pyriformis	134	17	stationary ⎭ without peptone		
Colpidium colpoda‡	59		50 day old culture	17°C	*Reich (1948)
C. colpoda‡	151		10 day old culture	17°C	*Reich (1948)
C. colpoda‡	191		1 day old culture	17°C	*Reich (1948)
C. colpoda‡	50–200		old growing culture		*Reich (1948)
C. campylum	200		24°		*Calkins and Summers (1941)
C. campylum	113		20°		*Calkins and Summers (1941)
Colpoda cucullus	113		trophic stage		Pigon (1959)
C. cucullus	13		cyst		Pigon (1959)
Colpoda sp.		35			Thimann and Commoner (1940)
Colpoda sp.	540–1210				Adolph (1929)
Paramecium caudatum	140–5600		various authors		*Calkins and Summers (1941)
P. caudatum	660–4500		various authors		*Danforth (1967)
P. caudatum	200–2000				Scholander *et al.* (1952)
P. aurelia	450–1500		various authors		*Danforth (1967)
P. aurelia	76–2289				Stewart (1964)
Bresslaua insidiatrix	80–400				Scholander *et al.* (1952)
Tillina magna	850–1700		trophic stage		Kawakami (1964)
T. magna	85		cyst		Kawakami (1964)

* Source, where original papers not consulted.   † Corrected.

‡ In some cases *Colpidium colpoda* may be synonymous with *Tetrahymena pyriformis* (= *T. geleii*).

6

Comparison of techniques for population estimation of soil and plant nematodes

M. OOSTENBRINK

Introduction

About 80 to 90% of all multicellular animals in the soil are nematodes. Their main significance in energy-flow cycles is connected with the widespread damage they cause as parasites of main crops; their role in formation and structuring of soil or disintegration of organic matter is small. Soils of farmland or under natural vegetation normally harbour a persistent, noxious nematocoenosis of several phytophagous species, varying in density and composition with the chances of introduction, plant cover, soil and climate. Population estimates for faunistic surveys, population studies and economic nematology require data on specific or even biotype level of nematodes and their food plants.

Sampling of nematodes from soil and from plant tissue is fairly easy because they are widespread at great densities and can stand up to handling. The characters of nematodes, on which techniques for extraction, concentration and enumeration are based, are: size and differential staining (hand sorting, direct microscopy, analysis of mixtures, identification), mechanical resistance (shredding, blender preparation), activity (wet-funnel methods, use of filters), shape (use of sieves and filters), specific weight (decanting, flotation, elutriation, centrifuging). The different techniques needed to evaluate the nematocoenosis in various situations are conveniently headed under: detecting nematodes *in situ*; extraction of all kinds of nematodes from soil; extraction of nematode cysts from soil; extraction of active nematodes from soil; extraction of active nematodes from plant tissue. The first three groups are used for special purposes; extraction of active nematodes from soil and from plant tissue are generally applied and therefore treated more extensively.

The purpose of this review is to compare the results obtained by the main techniques and evaluate techniques. For a detailed description of the methods I refer to the original authors or to the report on the Colloquium on Research Methods edited in *Progress in Soil Zoology* by Murphy (1962).

Methods for special purposes

Detecting nematodes *in situ*

Techniques are available for detecting nematodes in plant tissue by direct microscopy supported by differential staining, and detecting nematodes in soil by direct microscopy with help of special observation boxes. The methods are more qualitative than quantitative, but are nevertheless indispensable for diagnostic work and for ecological studies.

Extraction of all kinds of nematodes from soil

The only techniques for this purpose are direct microscopy of very small soil samples and centrifugal flotation in heavy liquids.

The first method (Seidenschwarz, 1923) is useful as a check on other methods and for special purposes, but is too cumbersome and time-consuming for general use. This holds too for the improvement by Minderman (1956), in which nematodes are differentially stained in the soil samples.

Centrifugal flotation (Caveness and Jensen, 1955) and its modifications also suffer from too many inconveniences to be accepted as a general technique. Also there is always a loss of nematodes in the process. But the possibility of having a universal method for the extraction of all kinds of nematodes, including cysts, other immobile stages and eggs is tempting and this should stimulate further attempts at improvement.

Extraction of nematode cysts from soil

Methods for the extraction of nematode (*Heterodera*) cysts from dry and from wet soil are well developed. The main problems are statistical, through the difficulty of counteracting irregular distribution. The choice of methods is governed mainly by cost.

Extraction of active nematodes from soil

Methods

Extraction of active nematodes from soil is a main aim of most nematode studies. Ten well-known methods, based on one or more characters mentioned in the 'Introduction' above and including simple and sophisticated techniques, are listed here, preceded by a figure for identification.

0 Direct microscopy (Seidenschwarz, 1923; Minderman, 1956)

0 Centrifugal flotation (Caveness and Jensen, 1955; Minderman, 1956; Jenkins, 1964)

I Baermann funnel (Baermann, 1917; Overgaard Nielsen, 1949; Andeson and Yanagihara, 1955)

II Direct cotton-wool tray (Oostenbrink, 1954a, 1960; Townshend, 1963)

0 Decantation + sieves (Cobb, 1918; Thorne, in: Goodey, 1951)

III Decantation + cotton-wool tray (Oostenbrink, 1960, 1964)

IV Decantation + sieves + filter (Christie and Perry, 1951; Tobar Jiménez, 1963)

V Inverted flask + sieves + filter (Seinhorst, 1955; Seinhorst, in: Murphy, 1962)

VI Split elutriator + sieves + filter (Seinhorst, 1965, 1962; Seinhorst, in: Murphy, 1962)

VII Funnel elutriator + sieves + filter (Oostenbrink, 1954a, 1960; Tarjan, Simanton and Russell, 1956; Vogel, in literature)

Comparative results

All data available on extraction efficiency by methods I–VII are collected and related to the funnel elutriator in Table 1. Experiments by different authors are indicated by letters A–F; the studies D and E refer to average results of groups of students of the International Post-graduate Nematology Courses at Wageningen in 1966 and 1967. Incidental data on comparison of two or three methods which do not fit in this table are collected in Table 2.

Full analysis by genera or species of the nematode populations obtained by different methods in the studies D, E, F of Table 1 are not recorded here, but they are available and are used in discussing selective extraction by some methods. The main phenomena concerning selective extraction are: (a) that soil or coarsely decanted soil placed directly in water for one or more days yields more saprozoic nematodes (probably due to hatching of eggs and breeding), more *Pratylenchus* larvae (emerging from excised root particles in the sample) and more *Heterodera* larvae (emerging from cysts); (b) that finely regulated elutriation causes more loss of large nematodes; (c) that sieving causes more loss of small nematodes.

Table 3 is based on detailed data of study E in Table 1 and indicates that variability of results between operators increases with decrease in percentage catch.

TABLE 1. Percentage efficiency of nematode extraction from soil.

Methods[1]	A 100 ml	A 100 ml	A 100 ml	B 50 g	C 50 ml	C 300 ml	D Recommended sample	E Recommended sample	F 50 ml
I	—	—	—	11 (3)[3]	20 (19)	6 (5)	—	—	—
II	—	—	—	71 (51)	—	—	—	—	116 (73)
III	—	—	—	100 (101)	—	—	82 (87)	85 (131)	133 (121)
IV	37	42	62	—	115 (115)	80 (82)	—	44 (63)	—
V	—	57	—	—	107 (89)	97 (98)	59 (57)	35 (48)	51 (51)
VI	55	80	—	—	—	—	50 (40)	90 (100)	82 (69)
VII	100	100	100	100 (100)	100 (100)	100 (100)	100 (100)	100 (100)	100 (100)

(Column group header: Experiments[2])

[1] I, Baermann funnel; II, Direct cotton-wool tray; III, Decantation + cotton-wool tray; IV, Decantation + sieves + filter; V, Inverted flask + sieves + filter; VI, Split elutriator + sieves + filter; VII, Funnel elutriator + sieves + filter.

[2] A, Malo (1960), four experiments; B, Oostenbrink (1960); C, Tobar Jiménez (1963); D, International Post-graduate Nematology Course, Wageningen, 1966; E, International Post-graduate Nematology Course, Wageningen, 1967; F, Unpublished data, October, 1967.

[3] Figures in parentheses are percentages relating to Tylenchida only.

TABLE 2. Comparison of incidental extraction techniques (percentages).

Stöckli, 1950
Direct microscopy (Stöckli, 1950)	100
Baermann funnel (Overgaard Nielsen, 1949)	25–86

Chapman, 1958
Baermann funnel (Anderson and Yanagihara, 1955)	87 (67–100)
Decantation + sieves + filter (Christie and Perry, 1951)	50 (32–77)
Inverted flask + sieves + filter (Seinhorst, 1955)	97 (81–100)

Townshend, 1963
Direct cotton-wool tray (Oostenbrink, 1954, 1960)	100[1]
Decantation + sieving (Thorne, in: Goodey, 1951)	54[1]

Stemerding, in literature
Funnel elutriator + sieves + filter (Oostenbrink, 1960)	100
Funnel elutriator + sieves + centrifugal flotation in $ZnSO_4$ 1·33 (Jenkins, 1964)	86

[1] Percentages relating to Tylenchida only.

TABLE 3. Comparison of five extraction techniques (experiment E in Table 1): variability within group of operators

Recovery[1] (%)	Variability		
	Results of all persons	Minus lowest	Highest four
III: 9 *persons, 50 ml*			
85 (131)	1: 5·2 (1: 4·0)	1: 2·7	1: 1·7
IV: 10 *persons, 300 ml*			
44 (63)	1: 5·5 (1: 5·0)	1: 5·2	1: 1·9
V: 8 *persons, 400 ml*			
35 (48)	1: 7·1 (1: 5·2)	1: 4·1	1: 2·6
VI: 8 *persons, 400 ml*			
90 (100)	1: 5·1 (1: 5·0)	1: 2·5	1: 1·4
VII: 9 *persons, 300 ml*			
100 (100)	1: 3·5 (1: 2·5)	1: 2·1	1: 1·3

[1] Figures in parentheses are percentages relating to Tylenchida only.

Evaluation

A short evaluation of the methods listed above under 'Extraction of active nematodes from soil', taking into account technical efficiency, cost and effort, is given as follows.

Direct microscopy is unpractical and *centrifugal flotation* suffers from too many inconveniences to be fit for general use. The same holds for *decantation plus sieving* as practised by earlier workers, because a dirty catch has to be evaluated, unless there is excessive handling with sieves which causes heavy loss of nematodes.

The *Baermann funnel* (I in Table 1) must not be used in its original shape for the extraction of soil nematodes, because the percentage recovery is below 20% of other methods. The loss is due to the depth of the soil layer in the funnel. Overgaard Nielsen (1949) probably got a better result because he used minute soil samples, and not because he applied additional heating and light.

The *direct cotton-wool tray* (II) is based on the same principle as the Baermann funnel but gives much better results, owing to the shallow layer of soil (50 g on a tray of 16 cm diameter, therefore 2–2½ mm in compact state) and owing to the special character of the filter which loosens the thin soil layer into an open structure. The method, however, is very selective: it catches excessive numbers of saprozoic nematodes because of breeding and hatching of eggs during the extraction, whereas the catch of tylenchids, especially of larger species, is poor (B and F in Table 1 and detailed data of analyses).

Decantation plus cotton-wool tray (III, normally called the 'decantation cotton-wool filter method') is the most economic and most efficient extraction method for soil samples up to 50 g or 50 ml. Decantation is coarse and therefore safe. A cotton-wool filter acts as a most efficient staple of sieves and at the same time as a filter. Some precautions must be taken that the final suspension is not contaminated by fine soil particles from the filter.

Decantation plus sieves plus filter (IV) decreases the chance of fine soil particles in the final suspension, but increases the loss of nematodes through the sieves. The method is rather time-consuming handwork and depends on personal skill.

The *inverted flask plus sieves plus filter* (V) is in fact a modification of IV, in which decantation is finely regulated and made less personal, but also less effective. The method is rather cumbersome and moderately efficient.

The *split elutriator plus sieves plus filter* (VI) has the advantage that it can deliver the nematodes in different categories by size. But it is cumbersome and time-consuming (30–40 minutes per sample) and expensive owing to breakage and difficult repair of the essential glass units. Efficiency is moderate.

The *funnel elutriator plus sieves plus filter* (VII), used as a standard here, is easy to handle and about as efficient as the decantation cotton-wool filter method (III). But it is not quantitative; in practical work an over-all efficiency of 70–90% may be reached. Losses of minute larvae (*Paratylenchus* in experiment F of Table 1) may be up to 30% when a staple of four sieves is used.

Extraction of active nematodes from plant tissue

Methods

Extraction of active nematodes from plant tissue may be realized with the Baermann funnel or its modifications. The method is inefficient, unless the nematodes are very active (e.g. *Ditylenchus* species) or unless the plant tissue is broken open by blender preparation (maceration according to Taylor and Loegering, 1953, or slight crushing of tissues according to Stemerding, 1963).

When treatment has to be extended, the water around the substrate has to be refreshed by stirring, washing or spraying. The essential part of Baermann's technique is not the funnel shape of the apparatus but the presence of moisture with sufficient oxygen (shallow water) around the material, so that the nematodes are concentrated in water as soon as they escape from the material (plant tissue, or sometimes soil with plant tissue). Improvements are made by: aerating or refreshing the water by a constant stream (Filipjev and Schuurmans Stekhoven, 1941); intermittent dipping (a system developed by the Swedish Plant Protection Institute); intermittent shaking of the material in water (Chapman, 1957; Minderman, 1956); spraying (Chapman, 1957; Goffart, 1959; Seinhorst, 1950). Young's incubation method (material in closed container, 1954) and Mountain's method (material in jars with liquid containing antibiotics through which air is passing, 1959) are also based on the principle of Baermann's method. The same holds for the four methods compared in Table 4: the unaerated beaker with water; the aerated beaker with water; the funnel-spray apparatus which combines dipping in Baermann funnels with Seinhorst's principle of spraying (Oostenbrink, 1954b, 1960) and the blender-cotton-wool tray (Stemerding, 1963) in which slight blender preparation is followed by placing the material on a cotton-wool tray.

Also soil with plant tissue or egg masses may require extraction as for plant tissue rather than for soil (de Guiran, 1966; de Maeseneer and van den Brande, 1960).

TABLE 4. Extraction of active nematodes from oak-leaf litter, samples of 5 g; extraction time 48 hours

Method	First experiment	Second experiment
Blender (5 sec): cotton-wool tray (Stemerding, 1963)	2,500	1,660
	1,410	1,540
	2,360	1,490
Funnel-spray apparatus (Oostenbrink, 1954b, 1960)	900	—
	890	—
	1,200	—
Aerated beaker with 150 ml water	410	320
	420	410
	360	460
Un-aerated beaker with 150 ml water	—	420
	—	400
	—	530

TABLE 5. Extraction of *Meloidogyne* larvae from soil with root debris and egg masses by six different methods (after de Guiran, 1966): number of nematodes per litre.

Elutriation of soil without preparation	0
Elutriation after 4 days of incubation of the soil in water	21
Elutriation after 16 days of incubation of the soil in water	3,120
Spraying + cellulose filter	4,474
Spraying + cotton-wool filter	17,726
Baermann funnel	32

Comparative results

Two sets of data on extraction efficiency of active nematodes from plant tissue (in soil or free from soil) are given in Tables 4 and 5, only to illustrate the influence of sample preparation, of water refreshing and of the filter used.

Evaluation

Table 5 indicates that the extraction of *Meloidogyne* from roots and egg masses in soil is stimulated by refreshing of the water by spraying, and also that cotton-wool filters give a higher catch than cellulose filters. Table 5 indicates that refreshing of the water by spraying is better than stirring of the water for extraction of nematodes from leaf litter. Stemerding's blender-cotton-wool tray, however, was more efficient than the funnel-spray treatment or other modifications of Baermann's technique. This is known to hold for extraction of nematodes from roots too.

Conclusion

Methods of detecting nematodes *in situ*, extraction of all kinds of nematodes from soil and extraction of cysts from soil are qualitatively evaluated only; these methods are used for special purposes and the choice between competitive techniques is mainly governed by specific aims or by economic considerations. Preparation of plant tissue and litter with a blender and use of cotton-wool filters appear to be advisable for extraction of active nematodes from these substrates (Tables 4, 5). Comparisons indicate that the decantation-cotton-wool tray (III) for small samples and the funnel elutriator (VII) for large samples can be recommended for extraction of active nematodes from soil (Tables 1, 3). A restriction has to be made for extraction of some large nematodes such as *Longidorus* and *Xiphinima* species which do not pass cotton-wool filters; for these nematodes larger-mesh sieves (175 μ) can be used and the catch should be placed on a finer screen (125 μ) of gauze or nylon in water for final extraction.

Acknowledgement

The author is indebted to Mr J.C. Rigg for his help in preparing the English text.

Bibliography

ANDERSON E.J. & YANAGIHARA (1955). A method for estimating numbers of motile nematodes in large numbers of soil samples. *Phytopathology*, **45**, 238.

BAERMANN G. (1917). Eine einfache Methode zur Auffindung von Anchylostomum-(Nematoden)-Larven in Erdproben. *Geneesk. Tijdschr. Ned.-Indië*, **57**, 131–137.

CAVENESS F.E. & JENSEN H.J. (1955). Modification of the centrifugal-flotation technique for the isolation and concentration of nematodes and their eggs from soil and plant tissue. *Proc. Helminth. Soc. Wash.* **22**, 87–89.

CHAPMAN R.A. (1957). The effect of aeration and temperature on the emergence of species of Pratylenchus from roots. *Pl. Dis. Reptr.* **41**, 836.

CHRISTIE J.R. & PERRY V.G. (1951). Removing nematodes from soil. *Proc. Helminth. Soc. Wash.* **18**, 106–108.

COBB N.A. (1918). *Estimating the nema population of soil.* 48 pp. Agric. tech. Circ. U.S. Dep. Agric. 1.

FILIPJEV I.N. & SCHUURMANS STEKHOVEN J.H. (1941). *A Manual of Agricultural Helminthology.* Brill, Leiden. 878 pp.

GOFFART H. (1959). Methoden zur Bodenuntersuchung auf nichtzystenbildende Nematoden. *NachrBl. dt. PflSchutzdienst (Berl.)*, **11**, 49–54.

GOODEY T. (1951). *Laboratory Methods for Work with Plant and Soil Nematodes.* 25 pp. Tech. Bull. Minist. Agric. Fish. Fd. 2.

GUIRAN G. DE (1966). Infestation actuelle et infestation potentielle du sol par les nématodes phytoparasites du genre Meloidogyne. *C. r. hebd. Séanc. Acad. Sci.* (Paris), **262**, 1754–1756.

JENKINS W.R. (1964). A rapid centrifugal–flotation technique for separating nematodes from soil. *Pl. Dis. Reptr.* **48**, 692.

MAESENEER J. DE & BRANDE J. VAN DEN (1960). Nieuwe inzichten in het onderzoek van grondmonsters op de aanwezigheid van Pratylenchus sp. *Meded. LandbHoogesch. OpzoekStns. (Gent)*, **25**, 1040–1046.

MALO S. (1960). Comparative efficiencies of three methods for extracting nematodes from root and soil samples. *Pl. Dis. Reptr.* **44**, 217–219.

MINDERMAN G. (1956). New techniques for counting and isolating free living nematodes from small soil samples and from oak forest litter. *Nematologica*, **1**, 216–226.

MOUNTAIN W.B. & PATRICK Z.A. (1959). The peach replant problem in Ontario. 7. The pathogenicity of *Pratylenchus penetrans* (Cobb, 1917). Filipjev & Stekhoven (1941). *Can. J. Bot.* **37**, 459–470.

MURPHY P.W. (1962). *Progress in Soil Zoology.* Butterworths, London. 398 pp.

OOSTENBRINK M. (1954a). Een doelmatige methode voor het toetsen van aaltjesbestrijdingsmiddelen in grond met Hoplolaimus uniformis als proefdier. *Meded. LandbHoogesch. OpzoekStns (Gent)*, **19**, 377–408.

OOSTENBRINK M. (1954b). Over de betekenis van vrijlevende wortelaaltjes in land- en tuinbouw. *Versl. Meded. plziektenk. Dienst (Wageningen)*, **124**, 197.

OOSTENBRINK M. (1960). Estimating nematode populations by some selected methods. In: *Nematology.* J.N. Sasser & W.R. Jenkins (eds.). 85–102.

OOSTENBRINK M. (1964). La ocurrencia y significación de los nematodos. *Conferencias Servicio Shell para el Agricultor* 1952–1962, 46–62.

OVERGAARD NIELSEN C. (1949). Studies on the soil microfauna. 2. The soil inhabiting nematodes. *Natura jusl.* **2**, 1–131.

SEIDENSCHWARZ L. (1923). Jahreszyklus freilebender Erdnematoden einer Tiroler Alpenwiese. *Arb. Zool. Inst. Univ. Innsbruck*, **1**, 37–71.

SEINHORST J.W. (1950). De betekenis van de toestand van de grond voor het optreden van aantasting door het stengelaaltje (*Ditylenchus dipsaci* (Kühn) Filipjev). *Tijdschr. PlZiekt.* **56**, 291–349.

SEINHORST J.W. (1955). Een eenvoudige methode voor het afscheiden van aaltjes uit grond. *Tijdschr. PlZiekt.* **61**, 188–190.

SEINHORST J.W. (1956). The quantitative extraction of nematodes from soil. *Nematologica,* **1,** 249–267.

SEINHORST J.W. (1962). Modifications of the elutriation method for extracting nematodes from soil. *Nematologica,* **8,** 117–128.

STEMERDING S. (1963). Een mixer-wattenfilter methode om vrijbeweeglijke endoparasitaire nematoden uit wortels te verzamelen. *Versl. Meded. plziektenk. Dienst (Wageningen),* **141,** 170–175.

STÖCKLI A. (1950). Ueber die quantitative Bestimmung der Bodennematoden. *Z. PflErnähr. Düng. Bodenk.* **51,** 1–22.

TARJAN A.C., SIMANTON W.A. & RUSSELL E.E. (1956). A labor-saving device for the collection of nematodes. *Phytopathology,* **46,** 641–644.

TAYLOR A.L. & LOEGERING W.Q. (1953). Nematodes associated with root lesions in Abaca. *Turrialba,* **3,** 8–13.

TOBAR JIMENEZ A. (1963). The behaviour of a soil population and some plant parasitic nematodes in the processes of extraction of five different methods. *Revta. ibér. Parasit.* **23,** 285–314.

TOWNSHEND J.L. (1963). A modification and evaluation of the apparatus for the Oosten-brink direct cottonwool filter extraction method. *Nematologica,* **9,** 106–110.

YOUNG T.W. (1954). An incubation method for collecting migratory endo-parasitic nematodes. *Pl. Dis. Reptr.* **38,** 794–795.

7

The Enchytraeids

F.B. O'CONNOR

Introduction

The Enchytraeidae are a group of microdrilid oligochaetes related to the Naididae and Tubificidae (Černosvitov, 1937). They are widely distributed, having been found in all parts of the world in habitats ranging from marine and freshwater littoral to sandy heaths. Almost any hand-full of soil can be expected to contain some enchytraeid worms. They are most abundant in acidic organic soils and are dependent upon a high moisture status. Drought seems to be the main factor limiting their distribution and abundance.

The taxonomy of the group has been ably revised by Nielsen and Christensen (1959, 1961, 1963) so that it is now possible to identify the majority of European species from live specimens under a light microscope. They are small; from 1–5 cm in length, and their bodily organization is simple (Fig. 1). The majority of species are bi-sexual hermaphrodites; mutual transference of sperm takes place and eggs are laid, singly or in groups of up to 35 in mucilagenous cocoons secreted by the clitellum. Some species are parthenogenetic (Christensen, 1961) and a few reproduce asexually by multiple fission (Bell, 1959; Christensen, 1959).

Contributions on the population ecology of the Enchytraeidae have been made recently by Nielsen (1954, 1955a, b), O'Connor (1957a, b, 1958) and Peachey (1959, 1962, 1963). Nielsen (1961) and O'Connor (1963) have made estimates of population metabolism in selected habitats. The reproductive biology of the group has not been extensively studied but Reynoldson (1939, 1943) and Ivleva (1953) have made valuable contributions on some species of economic importance. Springett (1963) has developed a useful large scale culture technique for soil dwelling species and Christensen (1956) has described a small scale method for observation of cocoon production in individual worms. O'Connor (1967) has reviewed the main body of information on the ecology of the group.

Chapter 7

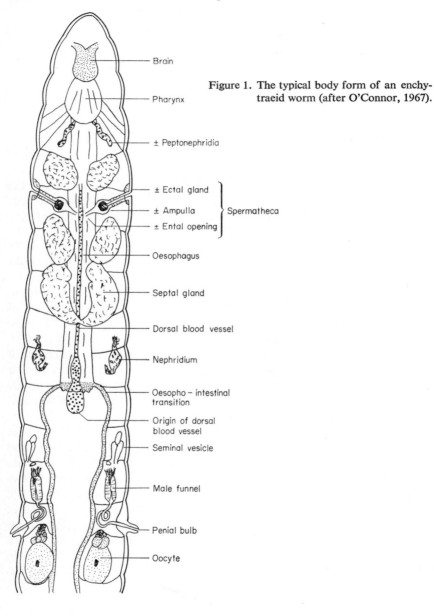

Figure 1. The typical body form of an enchytraeid worm (after O'Connor, 1967).

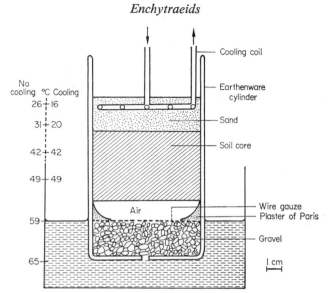

Figure 2. Nielsen's extraction method (after Nielsen, 1952–3, and also reproduced in O'Connor, 1967).

Population ecology

(a) Extraction from the soil

Two reliable methods exist for the quantitative removal of enchytraeids from soil. The principles of the methods and the construction of the two types of apparatus are illustrated in Fig. 2 and Fig. 3. In the first (Nielsen, 1952–3 the worms move upwards into the layer of cooled sand at the top of the core under the influence of temperature and moisture gradients. The form of the temperature gradient, with and without the cooling circuit in operation, is shown in Fig. 2. The importance of the cooling system is apparent. After about two hours all the worms will have moved into the sand layer which is then removed and the worms washed from it with a gentle stream of water.

The alternative method (O'Connor, 1955, 1962) relies on the downward movement of worms in response to the development of a temperature gradient over a period of about three hours. During this period the intensity of heating is gradually increased, by means of a variable resistance in series with the light bulbs, so that the temperatures shown in Fig. 3 are reached after about three

Chapter 7

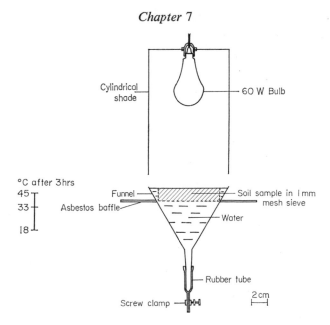

Figure 3. Wet funnel extraction method (after O'Connor, 1962 and also reproduced in O'Connor, 1967).

hours. The worms will, by then, have moved through the sieve and fallen to the bottom of the funnel. They can then be tapped off with a little water. Both methods are simple and the apparatus required is readily made. The number of units in use can be increased indefinitely. Nielsen's method is slightly more adaptable; the heating can be provided equally well from an electric element in the water bath or from a gas burner below it. The wet funnel method has the advantage of greater simplicity and extraction is a single process.

The efficiency of the two methods has been compared on a number of soil types (O'Connor, 1955, 1962; Peachey, 1962). There seems to be little difference in efficiency when used for extraction from mineral soils but the funnel method is approximately 1·5 times as efficient when used for extraction from highly organic soils such as peat or deep woodland humus.

(b) Removal of soil cores from the field

The enchytraeids are soft bodied, and some care is needed in taking soil cores, especially from heavy clay soils. Such soils tend to stick in ordinary cylindrical

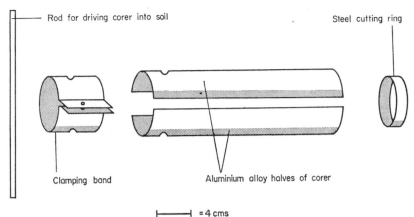

Rod for driving corer into soil

Steel cutting ring

Clamping band

Aluminium alloy halves of corer

|————| = 4 cms

Figure 4. The split corer (after O'Connor, 1957b).

corers and great force is needed to remove the core. This can impair extraction efficiency, both by damaging the worms and by compacting the soil. These risks can be avoided by the use of a longitudinally split corer (Fig. 4). This system has the great advantage that the corer insert can be opened to remove the core and that the core can very readily be divided into the natural divisions of the soil profile before extraction.

(c) Counting, identification and measurement of worms

The counting of extracted samples of Enchytraeidae is easy. They are readily visible under a low power ($\times 2$) bench lens and can be picked individually from a water-filled petri dish with a bulb pipette having an aperture of about 2 mm.

Identification of individuals to species is a more tedious process, requiring microscopical examination, and it is frequently practicable to examine only a proportion of the sample in this way. However, once the species content of the samples from a given site is known, it is often possible to recognize species from their setae. In general, it seems that few sites contain more than five species and of these only three will be present in large numbers. For example, a coniferous woodland soil in N. Wales contained three species: *Achaeta eiseni* (Vejdovsky) with no setae, *Marionina cambrensis* (O'Connor) with straight setae in bundles of two or three and *Lognettia cognettii* (Issel) with sigmoid setae. In situations such as this much time can be saved by making temporary mounts of several worms together in lactic acid or polyvinyl lactophenol and scanning under a microscope using a grid. These temporary preparations leave the clitellum and often the male funnels distinguishable so that it is possible to separate sexually mature from immature worms at a glance. Even with these aids the identification of extracted samples of enchytraeid worms remains a time consuming and exceedingly tedious task.

Similar temporary preparations can be used in order to measure size structure in the population. Worms preserved in alcohol and then mounted in polyvinyl lactophenol show a consistent relationship between length and fresh weight. Table 1 (after O'Connor, 1963) shows the length/weight relationship for three species from a coniferous woodland soil.

TABLE 1. The length/weight relationship of three species of Enchytraeidae.

Marionina cambrensis		*Cognettia cognettii*		*Achaeta eiseni*	
Mean length	Mean wt.	Mean length	Mean wt.	Mean length	Mean wt.
mm	μg	mm	μg	mm	μg
1·0	26	1·5	57	1·0	20
3·0	64	4·5	190	3·0	52
5·0	160	7·5	360	5·0	114
7·0	237			7·0	177

(d) Estimation of population density

The design of a sampling scheme for routine estimates of population density is influenced by the cost in man hours of the operations discussed above and also by several population characteristics. There is now sufficient information on the population characteristics of the Enchytraeidae (Nielsen, 1954, 1955a, b; O'Connor, 1957a, b, 1958; Peachey, 1959, 1962, 1963) for a definitive statement to be made of the requirements for an accurate estimate of density. These characteristics are discussed below under headings which indicate the kind of decisions to which their consideration leads.

(i) *General density levels—size of soil cores*

Most soils will contain between three and five species of enchytraeid and population densities will range from 10,000 to 300,000 per m². This is equivalent to from 20 to 600 individuals in a soil core of 20 cm² area. Densities in some areas may change by as much as 10 times during the season. Thus a core of around 20 cm² area will be sufficient to ensure that at least some worms are obtained on the majority of occasions and that the time required for counting and identification is not prohibitive. In areas where density is very high (2–300,000 per m²), the number of species present tends to be reduced so that counting and identification is not too time consuming even though numbers per soil core are high.

(ii) *Vertical distribution—depth of soil cores*

The Enchytraeidae are most abundant in the upper, well aerated, layers of the soil and do not usually occur below 10 cm depth. Table 2 shows the vertical distribution in a number of sites. As can be seen from the table, 90% of the

TABLE 2. Vertical distribution of Enchytraeidae in several sites

Site	% of total population					Author
	0–5	5–10	10–15	15–20	20–25 cm	
Coniferous woodland	78	21	1			O'Connor (1957b)
Sandy pasture	79	12	6	3	0	Nielsen (1955)
Arable land	51	16	12	8	10	Nielsen (1955)
Humid habitats	81	13	4	0	2	Nielsen (1955)

population is found in the top 10 cm of soil and a core of this depth will be adequate for most populations. During routine sampling, however, it is advisable to make periodic routine checks on vertical distribution.

In some soil profiles, where there is a distinct division into litter, humus and mineral profiles, different species are characteristically found to be most abundant in the different horizons (O'Connor, 1957b). In such cases there is considerable information to be gained from vertical stratification of the soil core before extraction. As mentioned on page 87 this is most easily accomplished using a split corer (Fig. 4).

(iii) *Horizontal distribution—number and position of sample units, analysis of sample results*

The most important consideration in the design of a sampling scheme for the Enchytraeidae is their horizontal distribution. It is apparent from studies by Nielsen (1954), O'Connor (1957b, 1967) and Peachey (1959, 1962, 1963) that density variation can occur both on small and large scales.

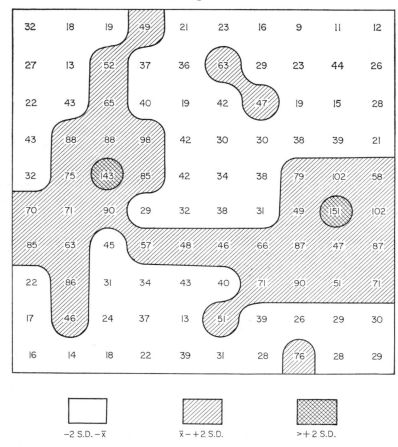

Figure 5. Distribution pattern of worms in a coniferous forest soil (after O'Connor, 1967).

Small scale variations in density, or micro-distribution patterns, have been described in a number of soils by means of complete enumerations of the population from small areas. A typical distribution map is shown in Fig. 5. There is a remarkable similarity in the density patterns recorded in sites ranging from sandy permanent pasture to waterlogged peat in spite of differences in species composition and general levels of density.

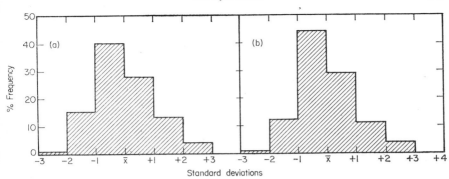

Figure 6. Frequency distribution of observations about sample means for random sampling and micro-distribution sampling (after O'Connor, 1967).

The significance of this density pattern for the design of sampling programmes is best considered in terms of the distribution of observations about sample means. Fig. 6b shows a histogram based on data from several micro-distribution studies by different authors and summarized by O'Connor (1967). The data are expressed in standard deviation units so that differences in general levels of density are ruled out. The form of the histogram is closely related to the distribution pattern in the soil: the proportions of the sample area occupied by different levels of density are represented in the histogram by the heights of the different blocks. Thus the relatively small area occupied by high densities is represented by a small number of observations between 3 and 4 S.D. above the mean while the general background density is represented by the large number of observations falling between the mean and − 1 S.D.

A consequence of this aggregative distribution pattern is that the standard deviation increases as mean density increases; this is apparent both for data from complete enumerations and from more widely dispersed random sample units. Fig. 7 shows this relationship for samples from a N. Wales coniferous woodland.

It is apparent that the density pattern is a characteristic of the family as a whole, differing little between species and sites. Thus this feature of enchytraeid populations provides a valuable method for checking the validity of population estimates based on dispersed random sample units. It is apparent that an adequate sample should reflect the aggregative distribution pattern, both in terms of the distribution of observations about the sample means and in terms of the relationship between standard deviation and size of the

Chapter 7

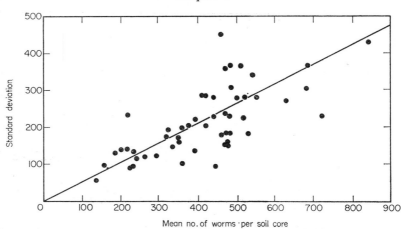

Figure 7. Relationship between standard deviation and mean size for random sampling from a coniferous forest soil (after O'Connor, 1957b; also reproduced in O'Connor, 1967).

sample mean. As an example, data from a number of random samples are expressed as a histogram showing the distribution of observations about sample means in Fig. 6a. The similarity of the two histograms is obvious. Strictly it would be desirable for every single sample taken during a routine census to be of sufficient size to replicate the characteristic distribution pattern. In practice, however, when the main interest is in examining population density trends, it is reasonable to lump samples from 2 or 3 occasions in order to examine their validity in the sense discussed above. In practice it is found that the minimum useful number of units per sample is 10 and that little increase in accuracy is obtained by increasing numbers beyond 25; sample variance becomes independent of the number of units per sample and the distribution of observations about the sample mean is not significantly different from that of the intensive micro-distribution sampling.

The aggregative distribution pattern also has an important bearing on the statistical analysis of results. The skewed distribution of observations about the sample means and the dependence of variance on the size of the mean invalidate many statistical tests, including the analysis of variance. This problem, however, is readily overcome by taking a logarithmic transformation of the original numbers per sample units and calculating log means. When the distribution of log observations about log means is plotted, it can be seen that the distribution is no longer skew (O'Connor, 1957b) and in addition the standard deviation and variance become independent of the size of the mean. Transformed data is thus suitable for the application of significance tests in which normal distribution and independence of mean and variance are accepted constraints.

The discussion of the most appropriate number of units per sample presented above is based on the assumption that no large scale gradients in the general level of population density occur within the sample area and that the sample units are to be distributed over a homogeneous area. In practice, it is necessary to examine the chosen sample area for the presence of large scale density gradients before embarking on a routine census. This is best done by taking soil cores at the intersections of a 2 m grid over the proposed sample area which will not generally need to be of more than 1 ha in area. It is then possible to examine the homogeneity of the area by sub-dividing it into a number of plots and comparing within and between plot variances by means of the analysis of variance. For this purpose it will be necessary to use logarithmic transformations of the original numbers per sample unit. When inspection of the data reveals no obvious heterogeneity an arbitrary sub-division of the sample area into 4 equal parts will be adequate but when obvious gradients exist some more appropriate divisions can be chosen. In any event, it will be preferable to reject areas which prove to be heterogeneous for routine sampling and even then it will be desirable to maintain a check by stratifying the routine samples between several sub-plots. In practice (O'Connor, 1957b) 4 sub-plots have proved adequate. In this way it is possible to perform periodic analyses of variance in order to test for possible trends towards non-homogeneity during the sample period.

(e) Rate of change in population density—frequency of sampling, duration of sample period

The population studies of Enchytraeidae carried out so far indicate that the dominant factors determining the rate of change in population density are temperature and moisture (Nielsen, 1955b; O'Connor, 1957b). In permanently moist situations density changes slowly in response to the seasonal variation in temperature but in situations liable to drought rapid changes in density can occur in response to changes in soil moisture status (Fig. 8). A basic pattern of monthly sample intervals is sufficient to reveal the general trend of population density and provided that weather conditions are adequately monitored, the natural buffering effect of the soil will permit rapid

Chapter 7

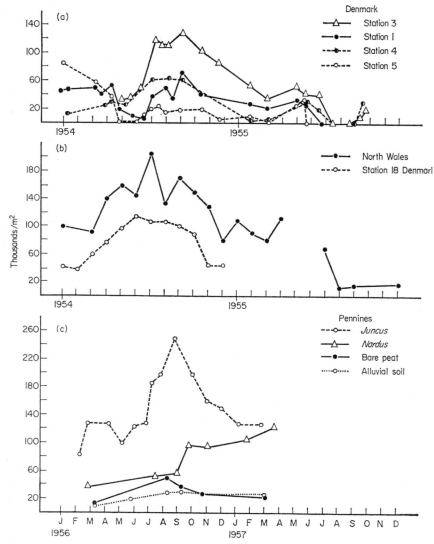

Figure 8. Population density of Enchytraeidae in (a) dry, (b) warm and wet, and (c) cold moorland (original data from Nielsen, 1955a, O'Connor, 1957b, and Peachey, 1963—reproduced in O'Connor, 1967).

density change to be anticipated so that the frequency of sampling can be increased in critical periods.

For population and production studies it will be necessary to sample over at least two seasons and preferably three. This is particularly so in populations in normally moist situations which have been dessimated by exceptional drought conditions. It is apparent (O'Connor, 1957b) that recovery from such an event can take at least one season, and possibly two. In situations where drought is the rule the population seems to be adapted to rapid recovery when the soil moisture status is restored (Nielsen, 1955a, b).

(f) Summary—final design of a routine sample scheme

On the basis of the discussion above, it is possible to arrive at an optimum plan for a routine sampling programme. The main features of such a plan are listed below.

(i) *Sample units*. Soil cores of 20 cm² area × 6–10 cm depth removed using a longitudinally split corer and subdivided into the natural divisions of the profile before extraction.

(ii) *Sample size* (*number of units per sample*). A minimum number of 10 units per sample is required to replicate the distribution pattern of individuals in the soil and, even then, samples from two or more occasions will have to be lumped for examination of distribution patterns. 25 units provide an optimum balance between accuracy and expenditure of effort in sampling.

(iii) *Distribution of sample units*. A stratified random sample is necessary in order to allow for periodic checks on the homogeneity of the study area.

(iv) *Frequency of sampling*. Monthly samples give a reasonable indication of the trend in population density and frequency can be increased in times of rapidly changing temperature and especially moisture conditions.

(v) *Size of study area*. The size of the study area required is easily calculated from size of the samples and the frequency and duration of the sampling. It is preferable not to sample at the same position more than once in a season. Using a sample site of 1 m² an area of 1 ha is sufficient to allow for a minimum of three sampling seasons at a rate of 25 units per month with sufficient area left over for micro-distribution studies and for increasing the frequency of sampling if necessary.

(g) Presentation of population data

The most appropriate method of presenting data on changes in population density and biomass will depend on the number of units per sample and on

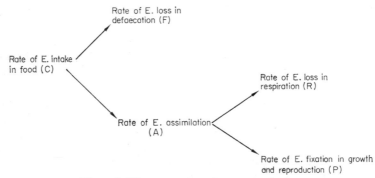

Figure 9. The parameters of an energy budget.

the purpose to which the data is to be put. When less than 25 units per sample are taken, it will be preferable to present data graphically as a running mean of two or three sampling occasions. In this way the number of units on which each point is based can be made sufficient to satisfy the requirements for validity discussed above. When 25 or more units are used, each point can stand by itself. If the aim is to examine the significance of changes in population density or to compare densities in different places it is necessary to use log transformed data. When the data is intended as a basis for calculation of production parameters the untransformed values will be more appropriate.

Production ecology

Studies of production ecology of the Enchytraeidae have been based on the simple descriptive relationship shown in Fig. 9. So far only the parameter R can be estimated with any confidence. However, it is probable that this accounts for a large proportion of the total energy intake by the population (Macfadyen, 1961) and, from the point of view of the functioning of the soil ecosystem, it is energy release and concomitant mineralization of substrate which is the most important parameter of the energy budget. Attention will, therefore, be concentrated on the measurement of the rate of energy release in respiration and such information as does exist for the other parameters will be used to indicate very approximate rates.

(a) Techniques for the measurement of the rate of oxygen consumption

The Enchytraeidae are simple animals, both structurally and physiologically. Their respiratory rate seems to be uninfluenced by temperature adaptation, diurnal or seasonal rhythms (O'Connor, 1963). They have a permanently moist body surface and are capable of surviving for long periods in isotonic aqueous media. Many of the smaller species must be dependent for movement upon the presence of waterfilms in the soil but individuals of more than about 1 cm in length burrow actively in the soil. Their respiratory rate remains unchanged when the oxygen tension of their surroundings is reduced by as much as 15%. These factors, taken together, make the measurement of respiratory rates of enchytraeid worms simple in comparison with many arthropods (Macfadyen, 1961; Phillipson and Watson, 1965) where the situation is complicated by diurnal and seasonal rhythms of activity. It is thus possible to use respirometric methods which are unsuitable for experiments of long duration. Nielsen (1961) and O'Connor (1963) have used the Cartesian Diver respirometer (Holter, 1943) (Fig. 10) for detailed studies of populations from well documented habitats. O'Connor (1967) has used both the diver and the Warburg respirometers for a more general study of levels of oxygen uptake in different genera and species. Of the two, the Warburg method is simpler and requires less rigorous control of laboratory conditions but it is not sufficiently accurate to deal with individual worms, especially of the smaller species. The diver, however, can be used for the smallest, newly hatched individuals of even the smallest species. In both systems the worms live in water in a closed vessel from which CO_2 is continuously removed and the O_2 tension gradually diminished by the activity of the worms. In spite of this, the rate of O_2 uptake by the animals remains the same for periods of up to 12 hours. By this time the O_2 tension in the vessel may have been reduced by as much as 15%. Since most enchytraeids live in the top few centimetres of the soil it is unlikely that they will encounter oxygen deficiencies greater than this.

(b) Oxygen consumption of common species of Enchytraeidae

The body of data available is presented in Fig. 11 from which the respiratory rates of the majority of common European species can be read. It is noticeable for all species, that the rate of oxygen uptake decreases with increasing body weight, reaching a minimum value at a weight of 1–2 mg. Above this weight the respiratory rate is independent of body size.

Fig. 12 shows a standard temperature correction curve based on several common species. The rate of oxygen uptake at 20°C is taken as 100% and values at lower temperatures expressed as a percentage of this. Using this curve, a simple percentage correction can be applied to Fig. 11 for different temperatures.

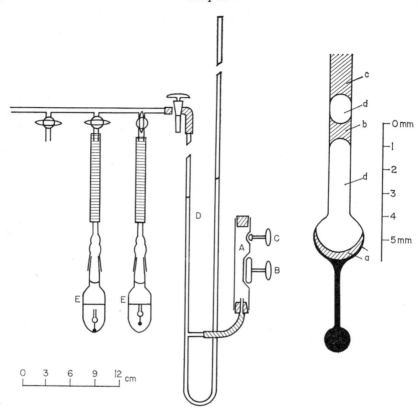

Figure 10. Schematical drawing of measuring apparatus and diver (after Holter, 1943). A–C pressure regulator. D manometer. E flotation vessels. a. bottom drop. b. oil seal, c. neck seal. d. air.

(c) Calculation of population metabolism

Used in conjunction with data on the population density, biomass and size structure, Figs. 11 and 12 permit the calculation of population metabolism for many common species. An example of such a calculation is given in Table 3 for a population of *Fridericia perrieri* (Vejdovsky) from a deciduous woodland site (Meathop Wood) in Lancashire for March, 1967.

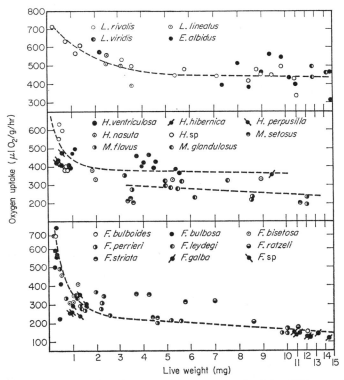

Figure 11. Oxygen uptake of common European species (after O'Connor, 1967).

Such calculations can be repeated for each sampling occasion and a picture of seasonal variation in metabolic activity produced. By averaging these values over a year a mean rate of O_2 uptake for the population can be calculated and readily converted to a total value of O_2 uptake during the year. This, in turn, can be converted to a value for energy release using the standard relationship (Brody, 1945) 1 l O_2 consumed = 4,800 cals released. Complete estimates of population metabolism have been made for Enchytraeidae only by Nielsen (1961) and O'Connor (1963) but it is apparent that no theoretical or practical difficulties exist in this kind of work. It is merely time consuming and rather tedious.

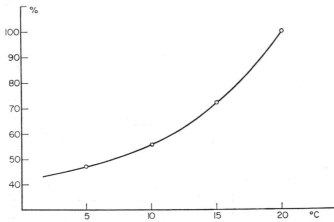

Figure 12. Temperature correction curve for respiration values cited in Figure 11 (after O'Connor, 1963).

TABLE 3. Calculation of population metabolism for a *Fridericia perrieri* at Methop Wood, March 1967. Field temperature 5 °C.

Mean body weight, mg	Nos/m^2	g/m^2	Oxygen consumption μl O/g/hr at 20 °C	at 5 °C	Population metabolism μl O$_2$/m^2/hr
2	1,650	3·3	250	112	371
5	1,050	5·25	210	94	496
8	300	2·4	175	79	189
					———
				Total	1,056

1056 μl O$_2$/m^2/hr = 709 ml/m^2/month = 3,400 cals/m^2/month

(d) The other parameters of the energy budget

While population metabolism can be estimated fairly easily for the Enchytraeidae, the measurement of the other parameters of an energy budget presents greater difficulties. The worms do not eat a recognizable substrate nor do they produce discrete faecal pellets. Most species simply burrow through the soil, ingesting it as they go, presumably utilizing organic matter, including bacteria and fungi from the soil. There are obvious difficulties in making quantitative estimates of food intake and digestion rates.

The rate of tissue production can be estimated, at least in theory from a knowledge of the rates of cocoon production, hatching, growth and mortality in the population. Culture experiments have been established in order to investigate these parameters but there remains considerable doubt about the relationship between laboratory and field conditions. At the present time it is possible to make only very approximate estimates of production from field data.

Nielsen (1955a, b, 1961) has described situations in which summer drought produces massive mortality and the population is virtually replaced each year by the mass hatching of cocoons already in the soil since before the drought. In such conditions, annual production of tissue must be at least equal to the maximum biomass obtained. In Nielsen's drought prone sites this amounts to between 4 and 8 g fresh weight per m². This would be equivalent to about 1 g dry weight and an annual production of 5,000 cals/m². It is interesting to compare this figure with the annual release of energy in population metabolism of about 40,000 cals/m² calculated from Nielsen's (1961) figures. A production/metabolism ratio of 1 : 8 does not seem unreasonable, especially since this is a minimum value for production to which must be added an unknown amount for continuous replacement in the population throughout the year. In permanently moist situations this marked period of replacement of the population does not occur. An analysis of the age structure of the population in a coniferous forest soil in N. Wales (O'Connor, 1958, 1963) showed that, although the production of juveniles was concentrated in the spring, there were no obvious changes in age structure during the rest of the year. From this evidence it can be deduced that mortality and replacement continue throughout the year. At present, there is no basis for estimating the rates of these processes.

(e) The influence of Enchytraeidae on total soil respiration

Since the Enchytraeidae are burrowing animals and process large amounts of soil, it is quite possible that their main importance will be in stimulating the activity of other components of the soil community. A respirometer has been designed in which it is hoped to measure this aspect of enchytraeid activity. The instrument is based on a micro oxygen electrode system supplied by Electronic Instruments Ltd., of Richmond, Surrey (model 19180). The electrode consists of platinum cathode and a calomel anode immersed in a molar solution of sodium carbonate in saturated potassium chloride. A polythene membrane is stretched over the tip of the platinum electrode. A potential of 0·65 v is applied to the cathode by means of an E.I.L.D48A polarizing unit and oxygen diffusing through the membrane produces a current on reduction at the cathode. The current produced has a linear relationship with the partial pressure of oxygen surrounding the electrode. The polarizing unit converts the current output to a voltage which, after suitable amplification, using an E.I.L. model 33B–2 Vibron Electrometer, can be recorded on a potentiometric chart recorder.

These basic components can be applied in a variety of ways to the measurement of respiratory rates. The application to respirometry of soil and soil animals is still under investigation but a preliminary design for a respirometer is shown in Fig. 13. The apparatus is assembled so that it can be used either as a continuous flow respirometer with electrodes situated at A and B (Fig. 13) or as a 'by-pass' system with only one electrode at B. In the former system the O_2 concentration of the air stream is monitored before and after passing over a respiring substrate while in the latter, air is passed alternatively through the by-pass and the respirometer chamber is compared periodically with the concentration of oxygen in air passed through the by-pass.

The sensitivity of the electrode system can be adjusted so that the voltage after amplication is 100 mv at the oxygen concentration in air (21% O_2) and a 1% change in oxygen concentration produces a 1 mv change in voltage. For use as a continuous flow respirometer it is convenient to adjust the flow rate so that 100 ml O_2 passes through the system in 1 hour. At 21% O_2, this corresponds to an air flow rate of 476·2 ml per hour. It is then apparent that a 1 mv change in voltage from the amplifier will correspond to an oxygen consumption of 1 ml per hour. Published values for soil respiration (Parkinson and Coups, 1963; Domsch, 1963) indicate rates of oxygen consumption varying between 0·5 and 5 ml per 100 g (dry weight) of soil. Thus the continuous flow

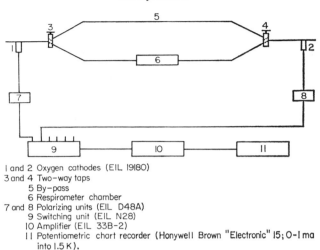

I and 2 Oxygen cathodes (EIL I9I80)
3 and 4 Two—way taps
    5 By—pass
    6 Respirometer chamber
7 and 8 Polarizing units (EIL D48A)
    9 Switching unit (EIL N28)
  IO Amplifier (EIL 33B-2)
  II Potentiometric chart recorder (Honywell Brown "Electronic" I5; O-I ma
    into I.5 K).

Figure 13. The oxygen cathode respirometer.

system would appear suitable for respirometry of 100 g soil samples. It is proposed to employ the system in order to investigate the effects of enchytraeid worm populations of different densities on soil respiration.

The alternative application of the system, as a by-pass respirometer, permits the detection of smaller changes in the rate of oxygen consumption but calibration is not so straightforward. In this system, air is passed alternately through the by-pass and the respirometer chamber. While the oxygen concentration of air passing through the by-pass is being monitored, an oxygen deficit is accumulating in the respirometer chamber. On switching the air flow to the respirometer, the voltage from the amplifier drops sharply and then rises slowly to its former value as the oxygen depleted air is flushed out of the respirometer chamber. Both the drop in voltage and the time required for it to return to its original value are proportional to the oxygen deficit accumulated in the respirometer chamber. Calibration of the by-pass respirometer is best done empirically by introducing known gas mixtures to the respirometer chamber and observing the drop in voltage and the recovery time when the respirometer chamber is flushed with the air stream. It is apparent that the voltage drop will be proportional to the time for which the respirometer chamber is closed as well as to the respiratory rate of the substrate. The duration of closure can be adjusted to suit various respiratory rates. Trials of the system with *Enchytraeus albidus* (Henle), living in water in the respirometer chamber, have given values for respiratory rate similar to those shown in Fig. 11.

Both applications of the apparatus are still under investigation but it is apparent that the system has the following advantages:—

(1) Great flexibility of the size of the respirometer chamber permits the investigation of substrates ranging from whole soil samples to individual animals.

(2) Carbon dioxide is not removed from the respirometer chamber so that the risk of influencing CO_2 sensitive respiratory systems is avoided.

(3) Provided that the air stream is brought to a constant temperature before passing over the electrode, the respiring substrate can be subjected to a variable temperature regime.

(4) Oxygen absorbed from the respirometer chamber by the respiring substrate is replaced, either continuously or intermittently depending on which system is in use.

(5) The system is readily adapted for automatic recording and is capable of providing records of respiratory rates over long periods.

General conclusions

Techniques for the identification and extraction of soil dwelling Enchytraeidae are now well established. The problems involved in the design and execution of a scheme for routine estimations of population density and biomass are well understood. It is possible, therefore, to make a definitive statement on methods for studying the population ecology of the group.

The construction of a complete energy budget for the enchytraeid populations is not yet possible. Of the parameters required, only population metabolism can be measured with any confidence. Work on tissue production, feeding biology and the influence of the worms on the level of activity in whole soils is at a preliminary stage. The latter, in particular, may well provide a fruitful line of research.

References

BELL A.W. (1959) *Enchytraeus fragmentosus*, a new species of naturally fragmenting oligochaete worm. *Science*, **129**, 1278.

BRODY S. (1945). *Bioenergetics and Growth*. 1,023 pp. New York.

ČERNOSVITOV L. (1937). System der enchytraeiden. *Bull. Ass. Russe Rech. Sci. Prague* 5,34, 263–295.

CHRISTENSEN B. (1956). Studies on Enchytraeidae. 6. Technique for culturing Enchytraeidae with notes on cocoon types. *Oikos*, **7**, 302–307.

CHRISTENSEN B. (1959). Asexual reproduction in the Enchytraeidae (Olig.). *Nature, Lond.* **184**, 1159–1160.

CHRISTENSEN B. (1961). Studies on cyto-taxonomy and reproduction in the Enchytraeidae with notes on parthenogenesis and polyploidy in the Animal Kingdom. *Hereditas,* **47**, 396–450.

CHRISTENSEN B. & O'CONNOR F.B. (1958). Pseudofertilisation in the genus *Lumbricillus* (Enchytraeidae). *Nature, Lond.* **181**, 1085–1086.

DOMSCH K.H. (1963). Die Messung von Abbaufolgen im Boden. In: *Soil Organisms.* J. Doeksen & J. van der Drift (eds.). 212–221. North Holland Publishing Company, Amsterdam.

HOLTER H. (1943). Technique of the Cartesian diver. *C.R. Lab. Carlsberg, Ser. Chim.* **24**, 399–478.

IVLEVA I.V. (1953). Growth and reproduction of the potworm (*Enchytraeus albidus*, Henle). *Zool. Zh.* **32**, 394–404.

MACFADYEN A. (1961). Metabolism of soil invertebrates in relation to soil fertility. *Ann. appl. Biol.* **49**, 215–218.

NIELSEN C.O. (1952–3). Studies on the Enchytraeidae. 1. A technique for extracting Enchytraeidae from soil samples *Oikos*, **4**, 187–196.

NIELSEN C.O. (1954). Studies on the Enchytraeidae. 3. The micro-distribution of Enchytraeidae. *Oikos*, **5**, 167–178

NIELSEN C.O. (1955a). Studies on the Enchytraeidae. 2. Field studies. *Nat. Jutland*, **4**, 1–58.

NIELSEN C.O. (1955b). Studies on the Enchytraeidae. 4. Factors causing seasonal fluctuations in numbers. *Oikos*, **6**, 153–159.

NIELSEN C.O. (1961). Respiratory metabolism of some populations of enchytraeid worms and free living nematodes. *Oikos*, **12**, 17–35.

NIELSEN C.O. & CHRISTENSEN B. (1959). The Enchytraeidae critical revision and taxonomy of European species. *Nat. Jutland*, **8–9**, 1–160.

NIELSEN C.O. & CHRISTENSEN B. (1961). Ibid. Supplement 1. *Nat. Jutland*, **10**, 1–23.

NIELSEN C.O. & CHRISTENSEN B. (1963). Ibid. Supplement 2. *Nat. Jutland*, **10**, 1–19.

O'CONNOR F.B. (1955). Extraction of enchytraeid worms from a coniferous forest soil. *Nature, Lond.* **175**, 815–816.

O'CONNOR F.B. (1957a). An ecological study of the enchytraeid worm population of a coniferous forest soil. Thesis (unpub.), University of Wales.

O'CONNOR F.B. (1957b). An ecological study of the Enchytraeid worm population of a coniferous forest soil. *Oikos*, **8**, 161–199.

O'CONNOR F.B. (1958). Age class composition and sexual maturity in the enchytraeid worm population of a coniferous forest soil. *Oikos*, **9**, 271–281.

O'CONNOR F.B. (1962). The extraction of Enchytraeidae from soil. In: *Progress in Soil Zoology.* P.W. Murphy (ed.). 279–285. Butterworths, London.

O'CONNOR F.B. (1963). Oxygen consumption and population metabolism of some populations of Enchytraeidae from North Wales. In: *Soil Organisms.* J. Doeksen & J. van der Drift (eds.). 32–48. North Holland Publishing Company, Amsterdam.

O'CONNOR F.B. (1967). The Enchytraeidae. In: *Soil Biology*. A. Burges & F. Raw (eds.). 212–257. Academic Press, London and New York.

PARKINSON D. & COUPS E. (1963). Microbial activity in a podzol. In: *Soil Organisms*. J. Doeksen & J. van der Drift (eds.). 167–175. North Holland Publishing Company Amsterdam.

PEACHEY J.E. (1959). Studies on the Enchytraeidae of moorland soils. Thesis (unpub.), University of Durham.

PEACHEY J.E. (1962). A comparison of two techniques for extracting Enchytraeidae from moorland soils. In: *Progress in Soil Zoology*. P.W. Murphy (ed.). 286–293. Butterworths, London.

PEACHEY, J.E. (1963). Studies on the Enchytraeidae (Oligochaeta) of moorland soils *Pedobiologica*, **2**, 81–95.

PHILLIPSON J. & WATSON J. (1965). Respiratory metabolism of the terrestrial isopod *Oniscus asellus* L. *Oikos*, **16**, 78–87.

REYNOLDSON T.B. (1939). On the life history and ecology of *Lumbricillus lineatus* (Mull.) (Oligochaeta). *Ann. appl. Biol.* **26**, 782–798.

REYNOLDSON T.B. (1943). A comparative account of the life cycles of *Lumbricillus lineatus* (Mull.) and *Enchytraeus albidus* (Henle) in relation to temperature. *Ann. appl. Biol.* **30**, 60–66.

SPRINGETT J.A. (1963). The distribution of three species of Enchytraeidae in different soils. In: *Soil Organisms*. J. Doeksen & J. van der Drift (eds.). 109–124. North Holland Publishing Company, Amsterdam.

8

Earthworms

J.E. SATCHELL

Introduction

The productivity of an organism can best be estimated if it is abundant; easily sampled in all stages of its life cycle; and possesses characters from which its growth rate may readily be determined. Lumbricidae are often both irregularly and sparsely distributed; concealed at depths from which they cannot easily be removed; and seasonally inactive when they cannot be sampled by expellants. Their external sexual characters develop and regress irregularly and they lack other morphological characters from which their age might be determined; their growth rate is irregular, subject to environmental change and periodically negative. They are therefore difficult subjects for production studies.

Very few estimates of earthworm production have so far been attempted and these are necessarily approximate. The ensuing sections describe methods of sampling, estimation of biomass, procedures used in a study of production in *Lumbricus terrestris*, and some aspects of population metabolism studies.

Methods of sampling field populations

Of the numerous methods used to sample earthworm populations, none are equally suitable for all species and all habitats. They are of three types; those that employ irritant solutions or an electric current to expel the worms from the ground; those in which soil samples are taken from the field and the worms subsequently removed from them; and trapping methods. The choice of method must be based on the distribution and behaviour of the species concerned, the type of soil it occupies, and logistic considerations.

(1) Extraction methods

EXPELLENT METHODS

(a) *Potassium permangate solution method*

Evans and Guild (1947), in their pioneering studies on earthworm populations of agricultural soils in Britain, used potassium permanganate solution
to expel worms from the soil, a method limited by the penetration of the
solution and the variable response of earthworms to it. The method has been
shown (Svendsen, 1955) to result for most species in considerable under-
estimates of population density and is not recommended.

(b) *Formaldehyde method*

The use of formaldehyde as an earthworm expellent was first described by
Raw (1959) and has been found particularly useful for studies of *Lumbricus
terrestris*. For species with a more horizontal and branching burrow system
which the solution cannot easily penetrate it is less suitable and it is also
unsuited for studies of aestivating species.

Concentration of solution. At Rothamsted, Raw tested solution strengths
of 25 ml of 40% formalin in 1 gallon of water, approximately a 0·55%
solution, and a quarter of this concentration, approximately 0·14%. The less
concentrated solution expelled less than 50% of the number of worms
expelled by the stronger solution in a population comprising mainly *Allolobophora chlorotica*, *A. caliginosa* and *A. rosea*. However, only seven pairs of
quadrats were sampled and the difference may not have been significant.

Subsequent tests at Merlewood compared ten 0·5 m² quadrats each treated
with four gallons (approximately 18 litres) of 0·165% formaldehyde solution
with ten quadrats treated with 0·275% solution with the following results:

	0·165%	0·275%
Adult and large immature *L. terrestris*	76	75
Small immature *L. terrestris*	275	270
Other species	589	632
Total	940	977

The differences were clearly not significant.

Concentrations of 0·275% and 0·55% were similarly compared on 23 pairs of
quadrats. The numbers of worms obtained were:

	0·275%	0·55%
Adult and large immature *L. terrestris*	420	429
Small immature *L. terrestris*	258	233
Other species	729	759
Total	1407	1421

These comparisons indicate that in the range 0·165% to 0·55% the concentration of solution is not critical. In subsequent studies at Merlewood, 0·275% solution (25 ml formalin in 2 gallons water) has been used.

Rate of application. The rate of application used by Raw (1959) was approximately 12 l/m² applied in two lots, the second about 20 minutes after the first. In a subsequent test on a woodland site at Merlewood, ten quadrats of 0·5 m² were treated with six applications each of 2 gallons (approximately 9 l, the capacity of a household bucket) of formaldehyde solution with 10 minute intervals between each application. Finally the top soil was removed and a further 2 gallons of solution was applied. The proportion of worms collected in each 10 minute interval is given in Table 1.

TABLE 1. The rate of expulsion of earthworms using successive applications of formaldehyde solution at 10 minute intervals.

Application	Number of worms (%) expelled between each application							Total Number Expelled
	1	2	3	4	5	6	7	
L. castaneus	79	19	1	0	1	0	1	85
A. chlorotica	74	18	3	2	1	1	0	407
L. terrestris	63	29	5	2	1	0	0	318
A. rosea	51	34	7	4	2	1	0	268
O. cyaneum	47	27	15	4	0	2	5	55
A. longa	41	41	10	0	3	3	3	39
A. caliginosa	40	37	16	5	1	1	0	141
All species	61	27	7			5		1313

It will be seen that the more active and/or surface-living species, e.g. *L. castaneus*, tend to emerge sooner than the deeper burrowing and/or less active species, e.g. *O. cyaneum*. However, taking all species together, 95% of the final catch had been expelled after three applications. As a better population estimate is likely to result from taking more samples than from striving for extreme accuracy with fewer samples, the procedure recommended is to apply three lots of 2 gallons (9 litres) of 0·275% solution at 10 minute intervals to each quadrat of 0·5 m².

Correction for soil temperature and moisture. Both soil temperature and soil moisture content have been found to have an effect on the number of worms expelled by formaldehyde solution. In studies on *L. terrestris* in English woodland, the maximum response to formaldehyde occurred at about 10·6°C and the numbers expelled increased linearly with soil moisture content to 38% and decreased at 44%. The correction term found appropriate (Lakhani and Satchell, 1970) was:

Estimated population =
observed popln. $\times \exp.\{0\cdot0075 \ (T - 10\cdot6)^2\} \times \exp.\{ -0\cdot0214 \ (M - 40)\}$
where T = soil temperature at 10 cm in °C
and M = % soil moisture content.

The model on which this formula is based is discussed by Skellam (1967). It should be stressed that the values of the constants in the correction term apply only to *Lumbricus terrestris* on the sites studied.

Efficiency of the method. Raw (1959) compared a formalin treatment using 0·55% formaldehyde solution at the rate of approximately 24 l/m² (2 gallons per 4 sq ft) with hand sorting to a depth of 20 cm (8 in). The latter method produced fewer *L. terrestris* than the formalin treatment so the comparison throws no light on the efficiency of formalin as an expellent for this species. The following results were obtained for other species:

TABLE 2. Comparison of extraction by formalin and hand sorting from Raw (1959).

| | Numbers per quadrat of 4 sq. ft (0·372 m²) | | | |
| | Arable soil | | Under grass | |
	Formalin	Hand sorting	Formalin	Hand sorting
L. castaneus	3·4	0·6	18·6	32·7
A. chlorotica	9·5	15·6	19·3	31·8
A. caliginosa	0·6	3·8	52·6	118·0
A. terrestris	0	0	3·1	7·7
A. rosea	2·4	7·5	11·3	55·2
O. cyaneum	4·8	6·3	3·0	8·2

At Merlewood, a one metre square quadrat and a 30 cm strip surrounding it, was treated with 0·275% formaldehyde solution in three applications each of 10 gallons (approximately 18 l/m²). The soil was then dug out to a depth of 100 cm and sorted by hand. From the square metre, 52 *L. terrestris* and 193 worms of other species were expelled by the formaldehyde; 6 *L. terrestris* and 77 of other species were recovered by digging.

The correction factors for soil temperature and moisture content led to an estimate of 58 *L. terrestris* and 216 other species for the population of this quadrat. Three more *L. terrestris* were found by digging than were accounted for by the correction factors, suggesting that for this species, if no further worms were missed by the hand sorting, the population would have been correctly estimated. Other species would have been under-estimated by about one-fifth. A formal assessment of the efficiency of the formalin method awaits the development of a less laborious method for comparison than digging and hand sorting.

HAND SORTING

Hand sorting, though laborious, is a widely used sampling method which is also frequently used for assessing the efficiency of other methods. Its efficiency has been tested by Raw (1960) and Nelson and Satchell (1962).

Raw carried out tests on a wet hill grassland soil with a thick mat of vegetation in which worms were confined to the top of the profile. Samples were sorted by hand and the vegetation and mineral soil were then re-examined separately. The vegetation was teased out in water to which a few drops of formalin had been added to make the worms wriggle. The mineral fraction was worked through a 2 mm sieve to break up soil crumbs and was then immersed in magnesium sulphate solution on which the worms floated. Of the total worms found of all species, 52% of the number and 84% by weight were found by hand sorting. Only 38% of the total number and 57% of the total weight of a small species, *Eiseniella tetraedra*, were found by the hand sorting. From similar tests on a heavy arable soil, 59% of the total number and 90% of the total weight was found by the hand method.

Nelson and Satchell tested the method by sorting prepared samples into which known numbers of worms had been introduced. Samples of rough grassland were freed of live worms by freezing and specimens of various sizes and species, including pigmented and unpigmented forms, were introduced. Of 924 worms introduced, 93% (99% by weight) were found by hand sorting but only 80% of immature pigmented *Allolobophora chlorotica* and 74% of *Lumbricus castaneus* were recovered. In general the method seemed to give a satisfactory recovery of individuals weighing more than 0·2 g live weight but not of smaller worms. For studies aimed at producing life tables, hand sorting would seem unlikely therefore to give a satisfactory recovery of the lowest age classes.

These tests were carried out with the operators knowing that their efficiency was being checked and may therefore have given an above-average recovery.

FLOTATION

The method used by Raw (1960) for extracting small species from soils unsuitable for hand sorting was based on an apparatus described by Salt and Hollick (1944). After preliminary washing, the organic material is floated off in magnesium sulphate solution in which the mineral soil sinks. The method has the advantage that it can be adapted to extract earthworm egg capsules and so enable all stages of the population to be estimated. Gerard (1967) states that in several tests this method extracted all the worms in the samples.

WET SIEVING

Washing soil with a jet of water through a series of sieves in the manner of Morris (1922) has been used by some authors, e.g. Krüger (1952), for collecting earthworms in studies concerned also with other faunal groups. There appears to be no published study of the efficiency of the method for extracting earthworms but its various disadvantages, e.g. damage to specimens and the long time and labour required to separate them from the residues, are listed by Ladell (1936). Raw (1960) found hand sorting much quicker than washing for both light and heavy soils. A method now under development (O'Connor, 1968) combines wet sieving with elution. The soil sample is broken up by water jets in the bottom of an inverted cone-shaped container, the dispersed sample flowing over an upper lip onto a submerged inclined sieve in another vessel. The soil passes through the sieve and the organic material moves downwards over its surface into a pipe leading to an elution unit. The flow of water through the pipe is such that the plant debris is carried up the elution column and the earthworms sink into a collecting chamber below it. The method has not yet been widely tested. The equipment is stated to process 4,000 cc of clay soil in one hour.

HEAT EXTRACTION

This method operates on the principle of the Baermann funnel (Baermann, 1917) and has been used for extracting small surface-living species from matted turf which would be difficult to sort by hand. The apparatus consisted of a plastic baby-bath, 55×45 cm, in which rested a wire gauze sieve 5 cm from the bottom of the bath. A battery of fourteen 60 watt light bulbs held in a hardboard sheet rested on the rim of the bath on two battens so that when a soil sample was placed in the sieve, the light bulbs were suspended about 2 cm above it. The bath was filled with water to about half-way up the soil sample which was heated by the bulbs for three hours. The heating unit, sieve and contents were then removed, and the worms were collected from the water immediately. The soil samples used in this apparatus were generally 20 × 20 cm × 10 cm deep.

The following results were obtained from a comparison of methods on a *Deschampsia flexuosa* site in an acid oak wood in which 90% of the earthworm population was *Bimastos eiseni*. Of 20 0·5 m² quadrats, ten were treated with formaldehyde solution (2 applications 0·275% +1 application 0·55%, soil temperature 10°C), as previously described. The soil and litter to a depth of 10 cm was removed from the remaining quadrats, that from three being sorted by hand while the remaining seven were heat extracted. The times taken to hand sort the three samples were 7¾, 9½ and 12½ hours. The mean numbers of worms per sample were: formalin 3·4; hand sorting 15; heat extraction 26.

TRAPPING

For very low population densities no satisfactory sampling methods have as yet been developed but baiting and trapping techniques have been found useful in comparative studies. Svendsen (1957) studied a *Calluna* moor from which fifty 20 cm³ soil cores yielded only one specimen of *Dendrobaena octaedra*. By examining at intervals 66 pieces of dung placed at the inter-sections of a 5 × 10 m grid he was able to analyse seasonal activity. The volume of the piece of dung was found to be important, pieces of 600, 300 and 150 ml yielding mean numbers of 58, 32, and 18 worms respectively over a period of 14 days. Boyd (1957) used prepared dung pats of 800 ml but these were disturbed by birds.

An alternative method in which the sorting may be carried out indoors is to prepare traps with a layer of ground dung over a layer of peat in a shallow earthenware plant pot. The traps are sunk to ground level and collected after 14 days for inspection in the laboratory. Thirty such traps laid out in acid oak woodland in Roudsea Wood National Nature Reserve yielded 84 *Bimastos eiseni* from an area of 120 m², an average of 0·7/m². Twenty soil cores from the same site treated by the heat extraction method gave an estimate of the *B. eiseni* population of 23·75±1·01/m². Although trapping methods are unlikely to yield accurate population estimates they can form a useful means of studying activity patterns where population densities are low.

(2) Determination of size and number of samples

Having decided on the appropriate extraction method, the investigator must then determine the size and number of sampling units required for population estimates of known accuracy.

Sampling unit area

A number of workers have followed Wilcke (1955) in selecting a sample unit area of 50×50 cm and a depth of 40 or 50 cm. Zicsi (1958) compared samples of this size with units of 25×25, 50×100 and 100×100 cm. He found that the number of worms per m^2 recovered by hand sorting decreased with increasing size of sample, indicating an increase in tediousness and hence of inefficiency in sorting the larger samples. In an arable soil the average number of worms estimated per m^2 from 50 units of each size were 25×25 84·5; 50×50 77·7; 50×100 62·8; 100×100 62·6. A tool for taking soil samples of 25×25 cm (1/16 m^2) to a depth of 20 cm is described by Zicsi for use in soils which are not too stony. Svendsen (1955) used a steel scoop of semicircular cross-section to cut cores 283 sq. cm (1/35 m^2) in area and 20 cm deep. The minimum sample area which will prove satisfactory will depend largely on the density of the populations to be sampled.

Number of sampling units

The number of samples which needs to be taken to ensure a population estimate of specified accuracy will depend on their size and the density and degree of aggregation of the population. This must be ascertained in the particular case by trial sampling but the following data from some typical temperate European sites (Table 3) indicate the sampling variance likely to be encountered. It will be found that a sampling error of less than 10% of the mean cannot generally be obtained without an excessive cost in sampling effort.

Biomass estimation

(1) Loss in weight during preservation

The British Museum (Anon., 1932) recommends preservation of earthworms in 70—90% alcohol after a preliminary fixing in 10% formaldehyde, but

TABLE 3. Numbers of samples required for estimating population means of given sampling error

	Numbers/0·5 m²									
1. A population of *L. terrestris* sampled by the formalin method	25	28	25	35	15	24	15	18	17	22

	Numbers/sq. ft									
2. A population of *A. caliginosa* sampled by an electrical method	7·7	1·0	5·4	6·0	4·2	7·0	3·8	11·8	3·7	7·0

3. A population of *L. castaneus* sampled by the permanganate method.

Numbers/sq. yd in a quadrat 8×8 yd

3, 2, 2, 4, 7, 7, 3, 3, 2, 1, 0, 0, 0, 1, 1,
5, 9, 1, 3, 3, 0, 0, 7, 0, 5, 4, 9, 9, 8, 8,
0, 2, 0, 0, 4, 12, 14, 6, 1, 1, 5, 3, 5, 0,
2, 7, 2, 0, 1, 14, 7, 4, 6, 4, 4, 2, 1, 6, 2,
0, 2, 6, 3, 1, 0.

Number of samples required to give a sampling error of the mean equal to

	5%	10%	20% of the mean
L. terrestris population	32	8	2
A. caliginosa population	103	26	7
L. castaneus population	358	89	23

ecologists more frequently use formalin as a preservative at concentrations of 4–10%. In both preservatives loss of weight occurs: at Merlewood, *L. terrestris* preserved in 70% alcohol were found to have lost 10% of their fresh weight in four days and Raw (1962) stated that the weight of worms preserved in 5% formalin is 25% less than the fresh weight. It is important to note that this weight loss varies with time as indicated by the following data from 50 *L. terrestris* preserved for 4 months.

TABLE 4. Loss in weight (%) of *L. terrestris* preserved in 4% formalin.

After	24 hours	7 days	1 month	4 months
Loss in weight	6·4	10·1	12·9	17·9

The worms used for these observations ranged in size from 1 g to 4 g. No relationship between size and the proportion of weight lost was detected. There is some evidence that greater weight losses may occur in some other species. This factor must therefore be taken into account when biomass estimates are to be based on samples of preserved worms.

(2) Conversion of fresh weight to dry weight

The analysis of the size class distribution of an earthworm population requires large numbers of worms to be weighed individually. If they are first collected into preservative this must be removed from the surface of the worm before weighing. The common practice of rolling the specimen on a cloth or filter paper is difficult to standardize and can be a source of serious error. It can be eliminated by determining the live weight/dry weight ratio and applying this to individual worm weights obtained from subsequent samples which have been collected into preservative, oven dried at 105°C and weighed from a dessicator. In studies where a life table can be determined solely from samples collected from the field it may be possible to work entirely on a dry weight basis but where the growth rate has to be computed in part from culture data the use of live weights is unavoidable. Regression analysis of the data from 44 specimens of *L. terrestris* which were weighed live, preserved for 4 days in 4% formalin, dried at 105% and reweighed, gave the relation:

$$1 \text{ g dry weight} \triangleq 5 \cdot 5 \text{ g live weight}$$

From a larger body of data, Lakhani and Satchell (1970) obtained the relation

$$1 \text{ g dry weight} \triangleq 6.37 \text{ g live weight}$$

In preparing estimates of the biomass of earthworm populations the weight of the gut contents should not be overlooked. A sample of 10 *L. terrestris* taken from the field and dissected had gut contents weighing 20% of the fresh weight of the whole worms but this value varied with the size of the worm. In a further sample of 30 worms selected to include a wide size range, the dry weight of the gut contents was on average 29% of the dry weight of the whole worm and ranged from 26% in worms of about 0·5 g live weight to 41% in worms of 7 g live weight. The regression of percentage weight of gut content on body size was significant at the 1% probability level. This result may have arisen from loss of gut material during dissection of the smaller worms or it may indicate that the gut is larger in proportion to the body in larger worms or that the larger worms ingest a higher proportion of mineral matter in their diet than the smaller worms. Bouché (1966a) has described a method of washing out the gut contents of large lumbricids employing a catheter and syringe and (1966b) a technique for measuring biovolume as an alternative to biomass.

Estimation of relative productivity

Relative productivity is defined as the ratio of the annual turnover of biomass to the average prevailing biomass. A first attempt to estimate it for two populations of *L. terrestris* has been made using the formaldehyde sampling technique and biometrical methods devised by J.G. Skellam and M.D.

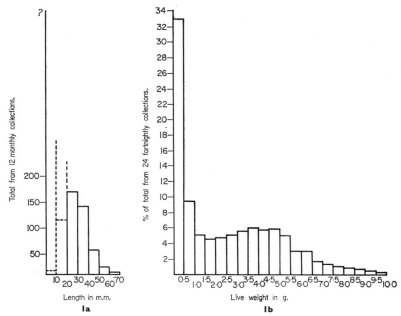

Figure 1a. Size class distribution of a population of *Eisenia rosea* (from Murchie, 1958).

Figure 1b. Size class distribution of a population of *Lumbricus terrestris* (Satchell, unpublished).

Mountford. The estimates are based on the size class structure of the populations, their growth rates and their mortality and survivorship curves. The methods described below are largely drawn from this work an account of which has been published elsewhere (Lakhani and Satchell, 1970).

(1) Size class structure

The size class structure of a population of *Allolobophora rosea* (calculated from Murchie, 1958) is shown in Figure 1a. Although the size classes are expressed in units of length which are less suitable for productivity studies than units of weight, the histogram illustrates the size class distribution typical of a species in which few individuals survive long after reaching the average maximum size. The data were obtained by hand sorting a soil sample of standard volume each month and expressing the numbers in each size group as a percentage of the total collected in the year. The smallest size class is under-represented because of the limitations of the hand sorting method. The histogram would otherwise have shown a continuously descending curve.

Figure 1b shows the size class structure of a population of *Lumbricus terrestris* sampled by the formaldehyde method (data from Lakhani and Satchell, in prep.). It illustrates the size class distribution of a species in which many individuals live on after the average maximum weight is attained. The histogram shows a hump at about 4·25 g which is interpreted as an accumulation of mature worms at about the average upper size limit for the species in the prevailing field conditions.

(2) Construction of growth curves

The construction of growth curves is straightforward for species in which the population overwinters as cocoons and the young worms hatch simultaneously and mature and die off with no overlap of generations. In these circumstances, provided that the population of newly hatched worms can be estimated adequately, the rate of growth in the field can be calculated directly from successive size class frequencies. Such straightforward cases are unfortunately rare, the more usual situation being that many sizes of worms are present in the population concurrently.

In *Lumbricus terrestris* the cocoons hatch mainly in the spring. This generates a peak in the size class distribution which can be followed for several months by determining the modal values of the weights of the worms collected at each sampling. Taking these values as the average weight of the worms on each sampling date, the growth curve can be constructed for about the first year. Thereafter the mode is obscured by the wide variation in growth rate of individual worms.

It may be possible in some studies to continue the growth curve by consideration not of the increase in average size with time, but of the average time taken to reach a given weight. If the growth curve is in a sharply ascending phase the modal value of the time class distribution at a certain weight will be more sharply defined than that of the weight class distribution at a certain time.

Beyond the point where these methods are applicable, the growth curve may be extended by use of culture data. The culture method used by the author for *L. terrestris* was to put newly hatched worms collected from the study population individually into nylon bags of worm free soil and leaf litter taken from the study site. These were buried on the site so that the litter was level with the soil surface. The worms were weighed initially and re-weighed and re-provisioned at monthly intervals. Because of the great variability in growth rate of individual worms a large number of specimens should be set up in culture; sixty were used in the work referred to.

This method keeps the worms at field moisture and temperature and mortality was much lower than when glass or earthenware containers were used. Food was doubtless more readily available than under field conditions and the culture worms grew faster and attained a greater maximum size than worms in the field population.

Before culture data can be used for constructing growth curves, it must be established that the general pattern of growth in culture is the same as that in the field. It may be possible to establish this for the beginning of the growth curve by comparing the growth in culture with that obtained from analysis of the size class distribution in the field. For the latter part of the growth curve this is likely to remain an unverifiable assumption. However, the sensitivity of the estimate of relative productivity to hypothetical departures of the culture growth curve from the field growth curve can be tested by appropriate models.

If it is established or assumed that the pattern of growth in culture is not materially different from that in the field, the next step is to adjust the rate of growth in culture to that in the field. This may be done by reference to
(1) the weight attained in the field at the last point where this can be deduced from the size class distribution
(2) the asymptotic weight attained in the field
(3) the weight attained in the culture at the time corresponding to weight (1)
(4) the asymptotic weight attained in the cultures.

Suppose (a) it had proved possible to construct a growth curve for the field population by analysing the movement of peaks in the size class distribution for 9 months and that the mean weight of the worms at 9 months was 1 g;
(b) the asymptotic weight in the field was 4 g (cf. Fig. 1b);
(c) the mean weight of the culture worms at 9 months was 2 g;
(d) the asymptotic weight of the culture worms was 8 g (cf. Fig. 2),
then the factor relating growth in the field to growth in the cultures would be:
$$(4-1)/(8-2)=0\cdot5$$
The growth curve from 9 months onwards could then be constructed by taking the weight increase in the field to be 0·5 times the weight increase in the cultures in the corresponding time interval.

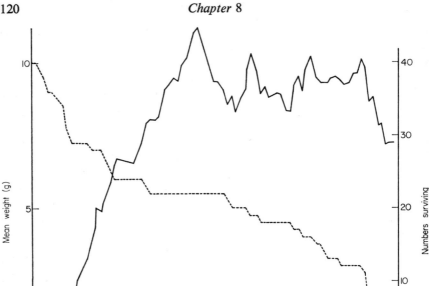

Figure 2. Growth increment and mortality (- - - -) of *L. terrestris* reared outdoors.

(3) Construction of survivorship curves

For species in which most of the worms hatch at about the same time, survivorshop curves may be constructed on the basis of the subsequent decline in the numbers of young worms. A size class histogram for such a species in the months after hatching will show an abundance of worms of very low weight followed by an absence of worms of slightly higher weight and then, in species with overlapping generations, the presence of worms of all higher weight classes. Assuming that the worms of higher weight are survivors of previous generations, data of this type can be conveniently tabulated in the following way:

	Apr.	May	June	July	Aug.	Sept.	Oct.	Nov.	Dec.	Jan.	Feb.	Mar.
1969 generation	250	224	199	146	130	188	110	97	92	80	63	46
Previous generations	200	192	211	186	179	159	165	195	167	175	180	189

In such a case the April worms might represent 250 of 1 month old and 200 of $13+25+37+ \ldots$ months old, the May worms 224 of 2 months old and 192 of $14+26+38+ \ldots$ months old, *et seq.*

The appropriate model for the survivorship curve will have to be constructed to suit the particular case and will be suggested by the shape of the survivorship curve of the worms in culture. It is likely to decline steeply at first when the worms have recently emerged from the cocoons, then to flatten out and finally to decline again as the worms become senile (cf. Fig. 2). This pattern can be described mathematically by the survivorship function

$$S(t) = \frac{1 - a\,t^k}{1 + bt}$$

where S = number surviving to age T
and t = time

a, the maximum age, and b and k are the parameters to be estimated. These may be determined by the method of maximum likelihood from the numbers counted at different times of the year and the age structure of the population.

(4) Relative productivity

The production by one generation of worms is the sum of the products of the numbers dying at all ages and their weights at death. The production per month is the sum of the product of the weight at death and the proportion of the generation dying in that month, i.e. the prevailing death rate. The relative productivity per month is the production per month expressed as a fraction of the prevailing biomass (standing crop).

Expressed mathematically, relative productivity on a monthly basis may be estimated by

$$\frac{\int_o^\infty \mu\,(a)\ S\,(a)\ W\,(a)\ d\,a}{\int_o^\infty S\,(a)\ W\,(a)\ d\,a}$$

where (a) = the death rate at age a,

and   S (a) = proportion of earthworms surviving to age a, both obtained from the survivorship curve,

and W (a) = average weight of the earthworms at age a obtained from the growth curve.

Population metabolism

The preceding section indicates procedures which may be found useful in estimating one component in the energy budget equation: Amount of food ingested = amount of tissue produced + amount lost in respiration + amount excreted.

(1) Ingestion

The amount of food ingested by earthworms depends upon (1) the rates of movement of material through the gut under different conditions, (2) the duration of these conditions. For species defaecating on the soil surface, estimates have been based on regular collections of worm casts (Evans, 1948) but for other species, authors (Barley, 1959; Satchell, 1967) have been able to suggest only approximate quantities based on estimates of the number of days in the year in which the worms might have been feeding. The rate of soil ingestion is known to depend on whether or not the worm is burrowing and to vary with environmental factors and the worm's physiological condition. A comprehensive analysis of these factors leading to a critical assessment of annual food intake has not yet been attempted.

The rate of movement of soil through the gut of *L. terrestris* has been studied by Parle (1963) using ^{32}P as H_3PO_4 as a tracer. Worms were kept singly in pots of soil containing pellets of labelled dung and peat and were dissected at intervals after the pellets had been taken. The results suggested that under the conditions of the experiment, food passed through the gut in about 20 hours. A similar result has been obtained (Satchell, unpublished) from worms kept initially in a light coloured subsoil then transferred to a dark topsoil.

(2) Excretion

From arguments based on laboratory estimates of nitrogen excretion rates of *L. terrestris* (Needham, 1957) and the seasonal pattern of soil temperature and biomass of a field population, Satchell (1963) suggested that the annual output of excreted nitrogen for this population might be of the order of 3·3 g/m². If we assume that this was in the form of 64% mucus, 16% urea and 20% ammonia as recorded by Needham for feeding worms, and that the calorific value of mucus is about 4·0 kcals/g and of urea is 2·5 kcals/g, the energy content of the non-faecal excreta produced by this population would appear to be of the order of 9 kcals/m²/year. The tissue production of this population is perhaps about 100 g live weight/m²/year. Assuming that (a) 1 g dry weight $=5 \cdot 5$ g live weight; (b) 30% of the dry weight is gut content; and (c) the calorific content of earthworm tissue is $4 \cdot 42$ kcals/g (French, Liscinsky and Miller, 1957), the calorific content of the tissue produced may be about 56 kcals/m²/year. Thus the energy content of the excreted metabolites may form a substantial component in the energy equation.

The measurement of the production of these metabolites has so far been attempted only under standard laboratory conditions and it is difficult to assess how the production of, for example, mucus in laboratory glassware relates to its production in undisturbed conditions in the soil. An alternative technique is being studied in which worms are 'labelled' by feeding them on organic matter containing N_{15}, transferring them to natural soil and measuring the rate of disappearance of the isotope from the tissues by analysis of specimens taken at successive time intervals.

(3) Respiration

A number of respiratory studies on lumbricids are currently in progress and it is premature to suggest which methods are most appropriate for the group. In three of the most recent studies the methods used were (1) a modified Lindgren method (Gromadska, 1962); (2) Warburg respirometry (Byzova, 1965); and (3) infra-red gas analysis (Bolton, 1967). As with population metabolism studies of other groups, the following factors are known to affect earthworm respiration:

Size. The respiratory rate per unit weight is higher for small specimens of a given species than for large specimens. This is well illustrated by Gromadska's (1962) data for *L. castaneus* and Krüger's (1952) data for *E. foetida* and is consistent with Konopacki's (1907), Davis and Slater's (1928) and Raffy's (1930) observations on *L. terrestris*. Byzova (1965) has shown that the dependence of respiratory rate on body size is marked for surface-active species (*Dendrobaena octaedra, Eiseniella tetraedra, L. castaneus, L. terrestris, O. lacteum*) but is almost lacking in subsurface-living unpigmented species (*A. caliginosa, A. rosea*).

Activity. Most workers in this field have sought to measure the respiratory rate of resting animals but Raffy (1930) gives some indication of the variation in metabolic rate with diurnal activity in *L. terrestris*. In observations continued for 48 hours, the maximum oxygen consumption rate was 13% higher than the minimum rate. Exposure to light, however, may double the respiratory rate (Davis and Slater, 1928). There are to date no published studies describing the activity pattern of field populations on which a realistic estimate of population metabolism could be based.

Ambient temperature. The respiration rate rises with increasing temperature. In poikilotherms generally, the magnitude of the temperature effect, described by the Arrhenius constant μ, has a value of 11,500 calories for a wide variety of respiratory processes. From experiments done at temperatures ranging from 9—27°C, a μ value of 11,040 calories has been found for *L. terrestris* (Pomerat and Zarrow, 1936) so the relationship between temperature and respiratory rate in earthworms appears to be similar to that reported for other poikilothermous animals. Acclimatization has been demonstrated in the tropical earthworm *Megascolex mauritii* in which the rate of oxygen consumption at 20°C was shown to be greater for winter animals than for summer animals by about one-third in large specimens to four times in small specimens (Saroja, 1961).

Soil atmosphere. Johnson (1942) and Byzova (1966) have demonstrated the dependence of respiratory rate on oxygen tension in experiments with mixtures of oxygen and nitrogen. Byzova found that soil dwelling species with a low respiratory rate are relatively insensitive to a drop in oxygen tension while surface-active species are most sensitive. The oxygen consumed in 10% O_2 expressed as a percentage of the consumption at atmospheric concentration was approximately, for *D. octaedra* 63%, *E. foetida* 70%, *L. terrestris* 93% and *A. caliginosa* and *A. rosea* 100%.

It is generally believed that carbon dioxide concentrations in the soil do not affect respiration greatly. *Eisenia foetida* shows no behavioural response to concentrations up to 25% by volume (Shiraishi, 1954), and Stephenson (1930) states that the presence of CO_2 up to a proportion of 50% has only a slight and reversible effect on earthworms. Extreme limits of CO_2 concentration in soils quoted by Russell (1950) range between 0·01 and 11·5%.

The respiratory data from various authors and the population estimates for English Lake District woodlands referred to earlier suggest that the annual oxygen consumption of these populations of *L. terrestris* may be of the order of 23 l/m² and the energy of respiration about 110 kcal/m² per annum. The energy of tissue production appears to be about half this amount and the energy excreted as metabolites perhaps one-twelfth.

References

ANON (1932). *Instructions for Collectors, No. 2, Worms.* British Museum, Natural History, London.

BAERMANN G. (1917). Eine einfache Methode zur Auffindung von Ankylostomum-(Nematoden)-Larven in Erdproben. *Meded. geneesk. Lab. Weltevr.* 41–47.

BARLEY K.P. (1959). The influence of earthworms on soil fertility. 1. Earthworm populations found in agricultural land near Adelaide. 2. Consumption of soil and organic matter by the earthworm *Allolobophora caliginosa* (Savigny). *Aust. J. agric. Res.* **10**, 171–185.

BOLTON P.J. (1967). The use of an infra-red gas analyser for studies on the respiratory metabolism of Lumbricidae. *Proc. IBP technical meeting on Methods of Study in Soil Zoology.* UNESCO, Paris.

BOUCHÉ M.B. (1966a). Sur un nouveau procédé d'obtentian de la vacuité artificielle du tube digestif des Lombricides. *Rev. Ecol. Biol. Sol. T. III,* **3**, 479–482.

BOUCHÉ M.B. (1966b). Application de la volumétrie à l'évaluation quantitative de la faune endogée. *Rev. Ecol. Biol. Sol. T. III,* **1**, 19–30.

BOYD J.M. (1957). Comparative aspects of the ecology of Lumbricidae on grazed and ungrazed natural maritime grassland. *Oikos,* **8**, (2), 107–121.

BYZOVA J.B. (1965). Comparative rate of respiration in some earthworms (*Lumbricidae, Oligochaeta*). *Rev. Ecol. Biol. Sol., T. II,* **2**, 207–216.

BYZOVA J.B. (1966). On the effect of oxygen tension upon the respiration rate in earthworms (*Lumbricidae, Oligochaeta*). *Rev. Ecol. Biol. Sol., T. III,* **2**, 273–276.

DAVIS J.B. & SLATER W.K. (1928). The anaerobic metabolism of the earthworm (*Lumbricus terrestris*). *Biochem. J.* **22**, 338–343.

EVANS A.C. (1948). Studies on the relationships between earthworms and soil fertility. II Some effects of earthworms on soil structure. *Ann. appl. Biol.* **35**, 1–13.

EVANS A.C. & GUILD W.J. McL. (1947). Studies on the relationship between earthworms and soil fertility. I. Biological studies in the field. *Ann. appl. Biol.* **34**, 307–330.

FRENCH C.E., LISCINSKY S.A. & MILLER D.R. (1957). Nutrient composition of earthworms. *J. Wildlife Mgmt,* **21**, 348.

GERARD B.M. (1967). Factors affecting earthworms in pastures. *J, Anim. Ecol.* **36**, 235–252.

GROMADSKA M. (1962). Changes in the respiration metabolism of *Lumbricus castaneus* Sav. under the influence of various constant and alternating temperatures. *Studia Soc. Sci. Torun. Sectio E (Zoologia),* **6**, No. 9, 179–189 and translation. 15pp.

JOHNSON M.L. (1942). The respiratory function of the haemoglobin of the earthworm. *J. exp. Biol.* **18**, 266–277.

KONOPACKI M.M. (1907). Oddychanie dzdzownie (Uber den Atmungsproze bei Regenwurmern). *Bull. Acad. Krakowie,* 357–431.

KRÜGER F. (1952). Uber die Beziehung des Sauerstoffverbrauchs zum Gewicht bei *Eisenia foetida* (Sav.) (*Annelides Oligochaeta*). *Z. vergl. Physiol.* **34**, 1–5.

KRÜGER W. (1952). Einfluss der Bodenbearbeitung auf die Tierwelt der Felder. *Z. Acker-u. Pflbau,* **95**, 261–302.

LADELL W.R.S. (1936). A new apparatus for separating insects and other arthropods from the soil. *Ann. appl. Biol.* **4**, 862–879.

LAKHANI K.H. & SATCHELL J.E. (1970). Production by *Lumbricus terrestris* (L.). *J. Anim. Ecol.* **39**, 473–492.

MORRIS H.M. (1922). On a method of separating insects and other arthropods from the soil. *Bull. ent. Res.* **13**, 197.

MURCHIE W.R. (1958). Biology of the oligochaete *Eisenia rosea* (Savigny) in an upland forest soil of Southern Michigan. *Amer. Mid. Nat.* **60**, 113–131.

NEEDHAM A.E. (1957). Components of nitrogenous excreta in the earthworms *Lumbricus terrestris* L. and *Eisenia foetida* (Savigny). *J. exp. Biol.* **37**, 425–466.

NELSON J.M. & SATCHELL J.E. (1962). The extraction of *Lumbricidae* from soil with special reference to the hand sorting method. In: *Progress in Soil Zoology*. P.W. Murphy (ed.). Butterworths, London. 294–299

O'CONNOR F.B. (1968). The estimation of earthworm populations with a note on a mechanical method of extraction. *Proc. IBP technical meeting on Methods of Study in Soil Zoology*. UNESCO, Paris.

PARLE J.N. (1963). A microbiological study of earthworm casts. *J. gen. Microbiol.* **31**, 13–22.

POMERAT C.M. & ZARROW M.X. (1936). The effect of temperature on the respiration of the earthworm. *Proc. Natn. Acad. Sci., U.S.A.* **22**, 270–273.

RAFFY A. (1930). La respiration des vers de terre dans l'eau. Action de la teneur en oxygène et de la lumière sur l'intensité de la respiration pendant l'immersion. *C. r. Seanc. Soc. Biol.* **105**, 862–864.

RAW F. (1959). Estimating earthworm populations by using formalin. *Nature, Lond.* **184**, 1661–1662.

RAW F. (1960). Earthworm population studies: a comparison of sampling methods. *Nature, Lond.* **187**, 257.

RAW F. (1962). Studies of earthworm populations in orchards. I. Leaf burial in apple orchards. *Ann. appl. Biol.* **50**, 389–404.

RUSSELL E.J. (1950). *Soil Conditions and Plant Growth*. Longman, London.

SALT G. & HOLLICK F.S.J. (1944) Studies of wireworm populations. I. A census of wireworms in pasture. *Ann. appl. Biol.* **31**, 52–64.

SAROJA K. (1961). Seasonal acclimatization of oxygen consumption to temperature in a tropical poikilotherm, the earthworm, *Megascolex mauritii. Nature, Lond.* **190**, 930–931.

SATCHELL J.E. (1963). Nitrogen turnover by a woodland population of *Lumbricus terrestris.* In: *Soil Organisms.* J. Doeksen & J. van der Drift (eds.). North Holland Publ. Co. Amsterdam. 60–66.

SATCHELL J.E. (1967). Lumbricidae. In: *Soil Biology.* A Burges & F. Raw (eds.). Academic, Press, London and New York. 259–322.

SHIRAISHI K. (1954). On the chemotaxis of the earthworm to CO_2. *Tohoku Imp. Univ. Sci. Rep. 4th Ser.* **20**, 356–361.

SKELLAM J.G. (1967). Productive processes in animal populations considered from the biometrical standpoint. *Proc. IBP Symposium on Secondary Productivity of Terrestrial Ecosystems.* K. Petrusewicz (ed.). Warsaw.

STEPHENSON J. (1930). *The Oligochaeta.* Oxford University Press.

SVENDSEN J.A. (1955). Earthworm population studies: a comparison of sampling methods. *Nature, Lond.* **175**, 864.

SVENDSEN J.A. (1957). The behaviour of lumbricids under moorland conditions. *J. Anim. Ecol.* **26**, 423–439.

WILCKE D.E. VON (1955). Kritische Bemerkungen und Vorschlage zur quantitativen Analyse des Regenwurmbesatzes bei zoologischen Bodenuntersuchungen. *PflErnahr. Dung.* **68,** 44–49.

ZICSI A. (1958). Determination of number and size of sampling unit for estimating lumbricid populations of arable soils. In: *Progress in Soil Ecology.* P.W. Murphy (ed.). Butterworths, London. 68–71.

9

Molluscs

P.F. NEWELL

Introduction

Terrestrial molluscs comprise those animals known as slugs and snails, and they live either on the surface or in the top 30 cm of soil. Most species are plant feeders and all lay eggs. In snails the size of the shell is dependent on the availability of calcium (Robinson, 1942; Ložek, 1962) and thin-shelled species, such as *Vitrina* sp. can be found even on acid soils. In slugs the shell is either very small and thin or entirely absent.

Most of the recent ecological research on molluscs is confined either to those groups that are of medical importance, usually freshwater hosts of schistosomiasis, or to marine species. For this reason I shall often refer to work done in these closely allied fields.

The majority of terrestrial molluscs are plant-feeders, and so are easy to assign to the first trophic level as herbivores. However, some are omnivores, some carnivores, and some even are cannibalistic. Thus, any consideration of energy flow through an ecosystem would have to be preceded by a detailed investigation into the biology of the feeding of the animal. The energy budget of even such a well-known land snail as the giant African land snail, *Achatina fulica*, is unknown. The digestive capabilities of molluscs have been reviewed by Owen (1966), the enzymes present in the gut may indicate possible energy sources.

Terrestrial molluscs have been studied in their own right for two main reasons. Firstly, because of the interest of European geneticists in the balanced polymorphism of *Cepaea nemoralis* (Harvey, 1964, for summary) and secondly, because of the world-wide interest in controlling the damage caused by slugs and snails to crops. In the tropics damage by *Achatina fulica* is of great importance (Mead, 1961). *Helix aspersa* is a pest of citrus orchards but in temperate climates snails are of secondary importance when compared with slugs, which damage such widely grown crops as winter wheat, potatoes and clover. In the absence of any standard procedure for accurately estimating numbers I propose to deal with 'research' and 'field methods' of estimating populations under separate headings. Research methods are those which are usually aimed at providing a reproducible estimate of known accuracy; simplicity and ease of operation are often of secondary importance. Field methods are frequently devised for estimating numbers in crops, many concentrate on trapping and with the inherent non-reproducibility of results, either tolerated or ignored. They are often simple to use, cheap to operate and can, in some cases, be related to more sophisticated methods by simple calibration.

Methods and apparatus for population and biomass estimation
Research methods

Animal numbers can sometimes be estimated by counting the total population within a defined area. This is a direct method and is useful for large and easily seen animals. For smaller animals similar direct methods are used, for example, by taking soil samples, extracting the animals and then multiplying the numbers caught by a factor to give the numbers per unit area or volume of soil. Indirect methods have been used for many years to estimate insect, bird and fish populations (Dahl, 1917; Lincoln, 1930; Jackson, 1936, 1939, 1940; Gillies, 1962) and rely on recapturing previously marked animals from the population.

Direct methods

The numbers of animals in populations of surface dwelling snails are far easier to estimate than those of subterranean molluscs. Snails are often counted by collecting all individuals within randomly distributed quadrats, or along transect lines. This method has frequently been used for sampling populations of *Cepaea nemoralis* (Goodhart, 1962), but Goodhart mentions that this method may be inadequate for the smaller snails. Using this method he estimated a population of 3·5 per m². Other workers (Bigot, 1963; Odum and Smalley, 1959) have estimated the biomass of snails by using 5×5 m² quadrats.

These methods are relatively simple to use on areas of low vegetation, such as grassland, but many molluscs, such as *Achatina fulica* are arboreal and cluster in numbers on the trunks of trees. Others, such as slugs, are subterranean, and with these sampling problems arise as with a variety of other soil animals.

Traditionally, slug numbers have been estimated by trapping methods. All of these are somewhat limited in accuracy as they cannot be related to total populations; the numbers that are caught are related to the numbers active. Similar criticisms are made of trapping methods for estimating populations of freshwater molluscs (Hairston *et al.*, 1958).

Quantitative estimates of the populations of terrestrial molluscs from soil samples (Morris, 1922 and 1927), also gave some measure of the vertical distribution although, unfortunately, neither the genera nor species were named. He took samples of soil and extracted the snails and slugs by hand sorting.

Soil animals are normally extracted in two main ways from soil samples either by a behavioural or by a mechanical method. For soil arthropods the typical behavioural method of extraction is by causing a dry zone to move slowly down a soil sample. This tends to drive animals into the moister and cooler parts of the sample and from there into a sampling tube. This is the method used when employing a Tüllgren funnel and it is obvious that the samples have to be processed immediately since the animals have to be alive or else they will not be extracted. Such behavioural methods are ineffective for both slugs and snails since the animals become desiccated and so are not extracted.

A mechanical method which has been used for many years is a soil washing system that is based on the different specific gravities of the soil and its fauna. This method (Salt and Hollick, 1944; Raw, 1955) involves separating the organic debris from the inorganic matter and fauna by washing with water through a series of sieves. The fauna is then separated from the inorganic matter by floating off the animals in magnesium sulphate solution of the correct specific gravity (1·17—1·20). This latter method has the advantage that the samples can be cold stored for some months before extraction as the animals are just as efficiently extracted dead as alive.

A modification of Salt and Hollick's method has been used for extracting slugs from soil samples (South, 1964) and it extracted eggs, adults and immature forms. Samples were collected and washed immediately. Because most slugs have no skeleton and readily decompose, then death during a period of storage may well greatly reduce the numbers extracted. Recent developments of soil washing techniques and some of the problems associated with them have been recently reviewed (Kevan, 1962; Murphy, 1962). It seems that soil washing methods are so laborious to use that most people employ them only as a last resort. This difficulty, combined with the need for a plentiful supply of water and good disposal facilities for the washed soil, means that even in the United Kingdom only the largest research establishments can afford to run them.

An apparatus for extracting Enchytraeidae in which the animals move into sand from soil in response to a moisture and temperature gradient (Nielsen, 1953) and recently reviewed by O'Connor (1966), has subsequently been used (Milne *et al.*, 1958) for extracting Tipulid larvae from turves. This method also drove out slugs from the soil but is of limited use (South, 1964). Slowly flooding the samples with cold water was more effective. This method, the cold water process (C.W.P.) was calibrated by soil washing and was found to be very efficient for *Agriolimax reticulatus* but slightly less efficient for other species. Thus, if this method is to be used for estimating total slug populations comparisons should always be made against a soil washing method to compensate for species differences and different soil types. The C.W.P. is of lower efficiency when used with arable samples than with turf samples. However, it is suggested (South, 1964) that thinner soil layers should be used whilst flooding arable samples.

Both soil washing and flooding methods have been used for studying the distribution of three species of slug and Hunter (1966) concludes that the C.W.P. flooding method, although less efficient, saves sufficient time and labour to justify its use. He has also made a comparison of the accuracy of sampling methods (Hunter, 1968) in the study of slug populations and concludes that the soil washing method is suitable for biomass estimation providing a correction is applied to account for the weight lost during extraction of the animals.

Snail populations have been sampled by counting the numbers within random quadrats. This method is probably satisfactory for surface-dwelling species in grassland habitats, but special methods have been evolved to separate those species that live in litter. Snails' shells have been separated from the litter by a flotation method (Vágvölgyi, 1953) but with this method over-estimates of the living population might be made as recently-dead animals would be extracted indistinguishably from live animals. Snail shells float in water because air becomes trapped in the shell whorls. Separation of plant material from snails was accomplished by soaking and agitating the sample of a solution of a molluscicide. The dead animals will sink to the bottom with the soil, leaving the plant material and empty shells floating on the top. This flotsam can then be decanted off and the remainder is then thoroughly dried. This process fills the shells with air so that when the sample is once again placed in water all the shells float, the soil sinking to the bottom. No estimate of biomass is given by this method.

If habitats are being sampled where there is little inorganic material present, as, for example, in woodland leaf litter, then the molluscs can be easily separated out by a simple flotation method (Williamson, 1959), which is similar to that of Vágvölgyi (1953). The leaves are washed in a water bath and the slugs and snails fall on to a piece of cloth. It is ineffective for situations where there is a lot of soil in the samples and so would be no good for extracting of snails which live in soil such as *Acanthinula aculeata*.

It has been suggested (Southward, 1966) that terrestrial ecologists might try to adapt the methods of extraction used by marine biologists for use with land soil samples. Such methods might involve a preliminary separation of the plant material by floating it off in water and then the separation of the molluscs from the soil by stirring with carbon tetrachloride (Birkett, 1957).

Indirect methods: population estimation by mark, release and recapture

In recent years indirect methods of population estimation have been developed. These have been widely used as a method for estimating populations of animals, especially insects (Jackson, 1940; Dowdeswell, Fisher and Ford 1949; Bailey, 1952; Leslie, Chitty and Chitty, 1953). The whole topic of marking small terrestrial animals has been reviewed (Dobson, 1962, and more recently, Southward, 1966) and most modern population estimations gain additional accuracy by marking recaptured animals again before releasing them a second time.

Snails can be dealt with in this way as their shells are easily marked with different coloured cellulose paints. This would make identification of individuals easy and also allow the application of several different marks on each individual. However, such marking methods must be of limited application for estimating populations of snails and slugs as many have distinct 'home-ranges' (Taylor, 1907, for *Helix aspersa*, *C. nemoralis*, *L. maximus* and *L. flavus*, Newell, 1965a and 1966 for *Agriolimax reticulatus*) and so released animals may well not mix randomly with the unmarked population; a fundamental prerequisite for accurate population estimates with this technique. It has recently been shown (Lomnicki, 1964) that releasing marked snails into a natural population does not produce any significant changes in the distribution of unmarked snails. Thus, providing mixing occurs of both marked and unmarked animals and that samples are equally likely to contain marked and unmarked animals, then marking methods can be used to estimate snail numbers.

Slugs cannot be marked with paints, except for the Testacellidae which have an external shell (Barnes and Stokes, 1951). Some slugs, for example *A. reticulatus*, can be marked by feeding them on a diet containing the stain Neutral Red (South, 1965; Newell, 1965a; Hunter, 1968). This marking method is only useful in marking those species that have a translucent skin, but has the advantage that it is harmless and easy to use.

Many animals can be labelled with radioisotopes and this has the advantage over more conventional methods that animals can be traced underground and also located at a distance (Webbe and Read, 1962). This latter advantage is important for field workers as it means that the marked animals can be located with a minimum of disturbance to the site. Slugs can be made to ingest radioisotopes (Fretter, 1952) and this technique has recently been used in Belgium (Francois *et al.*, 1965; Moens *et al.*, 1965) and in England (Newell, 1965a, b) for ecological studies. Population estimation by mark, release and recapture of radioactive animals has been attempted in England (Newell, 1965b). Adult *A. reticulatus* were marked with P^{32} in the laboratory and then released into the field. Recaptures were made at subsequent weekly intervals on the release area. This was an area of an undersown wheat stubble (27×20 yds) marked with canes into square yards. The slugs were released randomly over the whole area and traps placed at random within each square yard. At weekly intervals the area under each trap was carefully searched for slugs.

An inherent difficulty with this type of field experiment is the small proportion of animals recaptured but with slugs this proportion is large when compared with some animals (Gillies, 1962), such as mosquitoes, where only 1% of the marked animals were recaptured. With slugs 13·4% of the marked animals were recaptured 7 days after release, giving an estimated total population of 4236 $\pm$ 32% per acre (95% Fiducial limits). On subsequent recapture days the numbers of mature marked and unmarked slugs caught was much smaller and many dead slugs were seen. These figures are shown in more detail in Table 1.

However, Hunter (1968) has recorded very low recapture figures with Neutral Red marking which may be due to emigration from the plot or because the marks were lost, a difficulty also experienced by me when using this marking method.

TABLE 1. Summary of recapture figures and derived data from a field experiment.

Slugs released on 12th Nov.	19 & 20 Nov. (7 & 8 days)	26 Nov. (14 days)	3 Dec. (21 days)
Number of slugs caught on release area	567	472	538
Number of mature slugs caught	398	312	315
Number of immature slugs caught	165	160	225
Total number of marked slugs present, i.e. number released − number of marked light-coloured slugs caught	269	251	243
Numbers recaptured	36	22	8
Recapture efficiency	13·4%	8·75%	3·3%
Estimate of total population of release area	4236	5499	16312
95% Fiducial interval of this estimate	32·3% (5600–2870)	41·7% (7759–3239)	57·0% (25630–6990)

Field methods

Direct estimates

Most field methods for snails are similar to research methods, that is numbers are estimated by collecting snails from within randomly placed quadrats (Bigot, 1963, used 5×5 m quadrats) and then applying the appropriate correction for numbers per unit area (Table 2).

Similar methods have been tried for slugs (Bruel and Moens, 1958). In this paper they review the then current field techniques for population estimation and compare the efficiencies of sorting soil samples by hand with searching the soil surface. They found that surface collecting gave very variable results and tended to exaggerate the relative abundance of some of the species present. The numbers recorded varied with the time of year, climatic conditions and species. Hand sorting soil samples 0·3 m² was very slow and laborious but gave a more accurate estimate of the numbers of *A. hortensis* and *M. budapestensis* than surface counts.

Populations of *A. reticulatus* in grass leys have been studied in North America (Howitt, 1961) by taking surface counts within square yard quadrats but the method was abandoned because of the small numbers of slugs that were found. A comparison between the numbers of slugs caught by searching the surface of foot square quadrats and the numbers recorded after washing soil samples has been made (South, 1964; Hunter, 1968). More slugs were

TABLE 2.

From Bigot (1963).

Biomass in kilogrammes fresh weight per hectare of the vegetation and phytophagous invertebrates on a salt marsh in the Camargane.

Vegetation	Arthro-cnemetum glauci 10·750	Salicornietum fruticosae 37·750	Thero-brachypodion 18·000	'Sansouire' 24·370
Heteroptera	0·12	0·49	1·78	0·67
Coleoptera	0·9	1·4	20·53	3·17
Collembola	0·95	2·26	6·18	2·77
Myriapoda	0·05	0·01	0·19	0·08
Arachnida	0·64	0·98	3·1	1·37
Isopoda	0·5	9·4	18·2	4·74
Mollusca	1·4	3·88	4·2	3·15
Oligochaeta	—	—	101·15	24·27
Total	4·59	18·54	145·63	44·57

found by soil washing. Field trials with molluscicides have been evaluated by counting the numbers of dead animals seen on the surface within twenty randomized foot square quadrats (Gould, 1962). The inefficiencies of this method were recognized and he comments that the absence of an accurate sampling method inevitably makes it impossible for field workers to evaluate the results of their control measures.

Indirect methods

These methods are used only for those animals that cannot be reliably estimated by an easier direct method (see p. 129), and have been mainly used for slugs and often relying on the fact that slugs will come to rest in a dark and moist situation. Such 'refuge-traps' were compared (Miles, Wood and Thomas, 1931) and the numbers caught were then related to weather conditions. More recently it has been shown that the numbers of slugs caught by a metaldehyde and bran traps depends largely on the average night temperature and the length of time that the bait had been exposed (Webley, 1964). The higher the temperature the greater the numbers of slugs caught.

The numbers of slugs caught by different types of baits (Barnes and Weil, 1942) has been recorded with the object of improving control measures. These workers (Barnes and Weil, 1944, 1945; Barnes and Stokes, 1951; Barnes, 1953) have also tried to estimate population size by collecting all the animals seen throughout a standard collecting time. This method may give a reliable index of population size if a constant proportion of slugs are active on successive nights, on different sites and throughout the year. There is no evidence to suggest that any of these assumptions is correct. Such techniques have also been used for water and semi-aquatic snails (Hairston *et al.*, 1958) but they are not widely adopted.

A combination of 'refuge-traps' and baits (Thomas, 1944) has also been used to estimate slug populations. Standard baits were covered with black painted glass squares, 6×6 inches, and the numbers of slugs caught by them was recorded. He made it clear that one of the main uses of this method is to compare the efficiency of field control measures where, providing the conditions which govern slug activity are constant over the whole area, the proportion of the population active should be constant. He also attempted to calibrate trapping methods by determining the distance over which each trap attracted slugs; and so assessing its effective trapping area. He claimed that *A. reticulatus* were attracted from between $3\frac{1}{2}$ and $4\frac{1}{2}$ feet. This figure fits in well with later work (Newell, 1966) on the lengths of feeding excursions observed by cinematographic methods.

Other field methods rely on the amounts of standard foods removed during a standard period of time. Such figures have been used to forecast those fields in which damage to crops is likely to occur (Duthoit, 1961). A wide variety of foods has been used ranging from brassica plants (Miles, Wood and Thomas, 1931), carrots and beans (Bruel and Moens, 1958), wheat grains (Duthoit, 1961), clover leaves (Howitt, 1961) to brassica leaf discs (Trought and Heath, 1963).

Because these feeding and trapping methods are so simple to use, some attempt has been made to relate amounts of these standard foods eaten per night to the numbers of slugs (Newell, 1965a). The two methods used were the wheat grain method (Duthoit, 1961) and brassica leaf discs (Trought and Heath, 1963). The baits were covered with inverted seed pans and examined daily. The numbers scored for each trap for both of these methods gave different indices of slug numbers, one in terms of numbers of seeds damaged and the other of area of leaf eaten, but both of these depend on several important factors other than population size and temperature such as the abundance of natural food (e.g. clover) and the availability of shelter.

Leaf discs were cut with a spring-loaded cork borer (Heath *et al.*, 1964) from the leaves of Winter King cabbage and the amounts eaten per trap are compared with the wheat grain method in Table 3 (Newell, 1965a). The area of discs eaten by the slugs was measured by superimposing on them a 1 inch diameter wire mesh grid divided into 140 squares.

The figures for the mean percentage of leaf disc removed per day show far less variation than those for the wheat seeds and was approximately 1% per day. This seemed to show that this method was measuring a constant feature of the population and so it might be possible to relate the amount of food eaten per night to different sizes of population.

Energy flow

Introduction

Energy flow is a concept that has recently been applied to populations of soil animals and its value in placing groups of soil animals into their ecological perspective has been well summarized (Slobodkin, 1962; Macfadyen, 1963, 1964). The pioneer experiments were made in the United States (Lindeman, 1942; Odum, 1957) and such experiments have been continued with many

TABLE 3. Table showing the relationship between average night temperature and the amount of a standard food eaten by slugs.

Date	Average* night temp. °C	Numbers caught	Area of leaf disc eaten		Numbers of wheat seeds eaten	
			Amt. eaten in 1/140th of a 1″ dia.	% area of leaf disc	Mean	%
31 Oct.	6·75	—	—	—	—	—
1 Nov.	4·7	—	—	—	—	—
2 Nov.	3·7	—	21·22	3·789	0·3056	2·547
(Total for 3 days)						
3 Nov.	3·7	49	6·7	1·196	0·1944	1·620
4 Nov.	3·6	47	4·5	0·804	0·0833	0·694
5 Nov.	5·0	51	9·8	1·750	0·1111	0·926
6 Nov.	6·74	35	10·5	1·875	0·5278	4·398
7 Nov.	4·8	36	5·75	2·027	0·7500	6·250
8 Nov.	1·7	33	5·72	1·021	0·6667	5·556
			Total for 7 days = 9·50%		Total for 7 days = 10·19%	
			Total for 9 days + 11·56%		Total for 9 days = 21·99%	

* All temperatures were measured in short grass at soil surface with a mercury in steel circular thermograph.

TABLE 4. Table showing seasonal changes in species composition of the gut content of *Monacha cantiana* taken from a roadside bank, Reading, Berks. (Chatfield, unpubl.).

	March 1966	May 1966	June 1966	August 1966
Urtica dioca	++	+++	++++	++++
Lamium album	+	+++	++	+++
Anthriscus sylvestris	++++	++	++++	+
Heracleum sphondylium	—	—	+	—
Cirsium arvense	—	—	—	—
Grasses	+	++	+	++

++++ Eaten very frequently
+++   Eaten frequently
++   Eaten slightly
+   Little eaten
—   Rarely eaten

soil animals (see reviews by Hale, O'Connor, Satchell and Wallwork in Burges and Raw, 1966). Unfortunately, no such studies have yet been made with any terrestrial molluscs although intertidal species have received attention (Odum and Smalley, 1959; Paine, 1965).

Feeding in terrestrial molluscs

Slugs and snails usually feed on vegetable matter and, as herbivores, are assigned to the first trophic level. The analysis of plant remains in sheep faeces (Martin, 1955, 1964; Hercus, 1960) has enabled estimates to be made of the species composition of the diet of free grazing sheep. Such a study has also been made of the gut contents of slugs (Newell, unpublished) and it was found that they feed on most common weeds as well as many crops. Plants can be recognized by their crystalline cellular inclusions (as for example in the flower heads of *Chaenopodium album*), astrosclerides, as in leaves of *Capsella bursa-pastoris* and characteristic leaf hairs and epidermal cells in most other species. Frequently insect remains were found in the gut contents of slugs. It is not known whether these animals were being killed, and then eaten, or if the slugs were merely feeding on the dead bodies of the insects.

The snail, *Monacha cantiana*, will feed on most plants within its immediate habitat (Table 4) and in winter feeding ceases and the gut is often empty (Chatfield, personal communication). This is true, also, of many slugs.

TABLE 5. Table showing the range of plants and fungi eaten by *Arion ater* (Frömming, 1954).

Food	First experiment Amount eaten			Second experiment Amount eaten		
	sq. cm	gm	% body wt.	sq. cm	gm	% body wt.
LEAVES						
Allium porrum	3	0·18	2·8	6	0·4	3·6
Datura stramonium	74	2·90	15·5	71	3·1	18·0
Lamium album	11	0·13	4·2			
Nicotiana tabacum	12	0·45	19·6	16	0·7	18·5
Papaver somniferum	24	0·51	22·7			
Ranunculus lingua	14	0·30	19·4	13	0·32	18·8
Senicio vulgaris	22	0·60	8·0			
Sonchus asper	9	0·22	10·5	12	0·26	13·0
Sonchus oleraceus	19	0·34	22·6			
Kohlrabi	9	0·64	9·1			
Beetroot	4	0·12	2·7	5·5	0·2	3·3
Sugarbeet	8	0·27	5·8			
FUNGI						
Amanita mappa		0·11	2·6			
Amanita muscaria		0·8	26·6		2·6	16.4
Armillaria mucida		0·9	14·7		0·19	8·6
Boletus scaber		0·7	22·2			
Hebeloma crustuliniforme		0·48	6·8		0·7	10·4
Hydnum zonotum		0·17	3·1		0·04	1·1
Hypholoma epixanthum		0·27	6·2			
Scleroderma vulgaris		0·13	3·4			
FRUITS AND STORAGE ORGANS						
Ripe apple		0·3	9·0		0·55	8·9
Ripe pear		0·4	12·5		0·8	11·0
Kohlrabi 'root'		1·48	17·6			
Celery stems		1·1	15·8			
Potato		0·7	16·6			
Cucumber		12·4	42·4		5·8	44·5
Tomato		4·3	26·8		6·2	30·5

TABLE 6. Table showing the different parts of plants attacked by three different species of slug (Frömming, 1954).

Plant species	*Arion fasciatus*	*Arion hortensis*	*Arion intermedius*
Aegopodium podagria L.	++	+++	++
Alliaria officinalis Andr.	+ (1)	+++ (1)	++ (1)
Anemone nemorosa L.	+ (2)	++ (2)	—
Anemone ranunculoides L.	±	++ (1)	—
Anthericum liliago L.	+ (2)	+ (1)	+ (2)
Anthriscus silvestris Hoffm.	+ (3)	+++ (3)	+ (3)
Astragalus glycyphyllus L.	+	+++	+
Campanula rapunculoides L.	++ (2)	++ (1)	+ (3)
Cardamine pratensis L.	++ (3)	+++ (1)	+++ (1)
Chelidonium majus L.	+++ (2)	+++ (3)	+++ (2)
Cicuta virosa L.	+	++	+
Cirsium acaule L.	+++	+++	+++
Convolculus arvensis L.	+ (3)	++ (1)	+ (1)
Cynoglossum officinale L.	+ (1)	++ (1)	+ (2)
Datura stramonium L.	+++	+++	+++
Datura tatula L.	+	+++	++
Daucus carota L.	+	+++	++
Dianthus barbatus L.	+	++	++
Equisetum arvense L.	++ (1)	+++ (1)	++ (1)
Erysimum cheiranthoides L.	—	+	+
Eupatorium cannabinum L.	—	+	—
Euphorbia cyparissias L.	+ (1)	++ (1)	++ (1)
Euphorbia helioscopia L.	+	++	+
Gagea lutea Schult.	++ (2)	++ (1)	++ (3)
Galeobdolon luteum Huds.	++ (3)	+++ (1)	++ (1)
Glechoma hederacea L.	+++ (1)	+++ (1)	+++ (1)
Lamium album L.	++ (1)	+++ (1)	++ (2)
Lathyrus pratensis L.	+ (2)	++ (3)	+ (2)
Linum usitatissimum L.	++ (3)	++ (1)	++ (3)
Majanthemum bifolium Schm.	+	+	—
Malachium aquaticum Fr.	++ (3)	+++ (1)	++ (1)
Pachysandro terminalis S. & Z.	—	+	±
Polygonatum multiflorum All.	+	+	+
Ranunculus acer L.	++ (2)	++ (3)	++ (3)
Raphanus raphanistrum L.	+++ (1)	+++ (1)	+++ (1)
Rheum rhaponticum L.	±	++	±
Senecio vernalis W. & K.	++ (3)	++ (1)	+ (1)
Sisymbrium officinale Scop.	++	+++	++
Spinacia oleracea L.	++	+++	++
Tragapogon pratensis L.	++ (3)	++ (1)	++ (3)

The numbers and plus and minus signs give information about the parts of the plants eaten and the severity of attack.

(1) Only petals.   (2) Young leaves.   (3) Stems and leaves.

— No feeding.   + Slight amount of feeding.   ++ Many holes or feeding areas. +++ More intense feeding up to the total destruction of the plant.

TABLE 7. Fungi attacked by two species of slug, in % body weights
(Frömming, 1953).

Species of fungus	*Limax maximus*	*Agriolimax reticulatus*
Amanita junquillea Quel.	27·2% (All)	9·5% (Cap)
Amanita mappa Batsch	30·5% (,,)	12·4% (,,)
Amanita muscaria L.	34·0% (,,)	11·3% (Epidermis)
Amanita phalloides Fr.	36·6% (,,)	10·6% (Cap)
Amanita rubescens Fr.	33·7% (,,)	7·5% (,,)
Amanita verna Bull.	24·8% (,,)	11·2% (,,)
Armillaria mellea Vahl	25·5% (,,)	12·8% (Gills)
Armillaria mucida Fr.	26·8% (,,)	Not attacked
Boletus scaber Fr.	10·3% (,,)	22·2% (Cap)
Cantharellus aurantiacus W.	3·4% (Gills)	6·8% (.,)
Clitocybe laccata amethystina Bolt.	5·6% (All)	Not attacked
Flammula sapinea Fr.	10·5% (Cap)	,, ,,
Hebeloma crustuliniforme Fr.	20·6% (All)	,, ,,
Hydnum zonotum L.	7·6% (Cap)	4·3% (Cap)
Hypholoma epixanthum Fr.	8·6% (,,)	Not attacked
Hypholoma sublateritium Fr.	15·2% (,.)	,, ,,
Lactarius torminosus Schäff.	14·0% (,,)	,, ,,
Lepiota procera Scop.	10·5% (,,)	,, ,,
Pluteus cervinus Fr.	22·0% (All)	8·7% (Cap)
Psalliota arvensis Schäff.	28·8% (,,)	20·1% (All)
Polyporus squammosus Huds.	17·5% (Stem)	Not attacked
Russula fragilis Pers.	18·0% (All)	16·2% (All)
Tricholoma georgii Clus.	12·6% (Cap)	4·6% (Gills)

Much information has been collected by Frömming (1953, 1958, 1962) about the food preferences of European terrestrial molluscs. Not only has he investigated the species of plants eaten by molluscs (Table 5) but he has also shown that different species attack different parts of the plant (Table 6).

Frömming (1953) also showed that some species feed preferentially on fungi. The amounts of various species of fungi that are eaten by two species of slug are shown in Table 7 expressed as percentages of their body weights. Because of the lack of understory plants in coniferous forests those species that can feed on fungi tend to predominate, for example, *Arion ater rufus*, *A. subfuscus* and *A. fasciatus* (Frömming, 1958). Their faeces are, in turn, an additional source of food for fungi and bacteria.

The animals secrete large quantities of mucus when they crawl and egest partially digested plant material as faeces. The gross chemical changes associated with ingestion of oak leaves by the caddis-fly larvae, *Enoicyla*

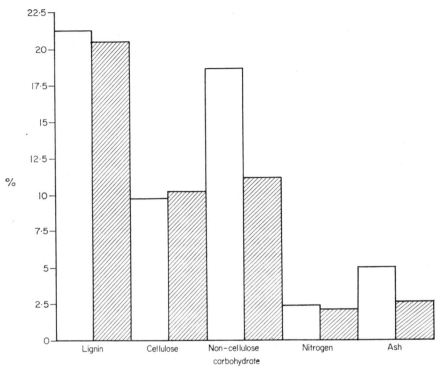

Figure 1. Chemical constituents of oak leaves in %, oven dried leaves (white columns) and faecal pellets (hatched columns). Drift and Witkamp (1960).

pusilla, has been shown to be mainly associated with large changes in the non-cellulose carbohydrate content (Fig. 1). Not only is the chemical composition altered but also the physical nature of the ingested food. It has been estimated that *Enoicyla pusilla* larvae (Drift and Witkamp, 1960) produce a two-fold increase in volume of ingested food and a fifteenfold increase in surface area. This alteration in physical properties produces better aeration and increased water holding capacity. These events, in turn, profoundly influence the numbers and species of bacteria and fungi that are able to assimilate the litter after it has been processed by the caddis fly larvae. The increased microbial activity has been correlated with increased respiration, measured as CO_2 evolved, within the faecal pellets when compared with whole leaves.

TABLE 8.

	Pomatias elegans	Helix pomatia	Cepaea nemoralis
Density/m²	80	1	1
Authority	Frömming, 1958	Frömming, 1958	Schnetter, 1950 (quoted by Frömming, 1958)
Wt. food consumed/ day/animal	0·15–0·18 g	4–6 g	0·8–1·0 g
Food consumption of popl./m²/day	12–14 g	4–6 g	0·8–1·0 g
Food consumption of popl./m²/year	4·32–5·18 kg	1·44–2·16 kg	0·29–0·36 kg

Oak coppice litter production was measured by Drift and Witkamp (1960) to be between 0·325 and 0·350 kg/m²/year.

It seems reasonable to assume that similar changes could be measured within the food and faeces of all terrestrial molluscs. Estimates of the food consumption of three species of mollusc are shown in Table 8. Although no measurements have been made it is widely held that land molluscs have an important role in humus building on some sites (Frömming, 1958).

Although most molluscs are herbivores some important groups are omnivores or carnivores. Many, if kept in crowded conditions are cannabalistic, as, for example, the snails *Oxychilis* sp. (Taylor, 1914) and the slugs *Arion hortensis*, *Milax budapestensis*, and *Agriolimax reticulatus*. The importance of this behaviour in the field is probably small. The slug *Arion ater* will feed on any meat or carcass and it is often seen in the road feeding on a dead hedgehog or bird that has been killed by a car. Slugs of the genus *Testacella* will feed on other slugs and on earthworms, often hunting worms down their burrows (Barnes and Stokes, 1951).

Predation in molluscs has been well reviewed (Mead, 1961) and in Britain both slugs and snails form an important part of the diet of some birds (*Turdus* sp.) and also hedgehogs (Dimelow, 1963). Some flies also feed on molluscs (Berg, 1953, 1961; Knutson, 1962; Knutson, Stephenson and Berg, 1965) and in this country the beetle, *Carabus violaceus*, will kill slugs (Moore, 1934). In the tropics the beetle larvae of *Lamprophorus tenebrosus* will attack the snail *Achatina fulica*.

TABLE 9.

Species	Oxygen used in 24 hr period whilst		Estimated oxygen requirement for 24 hr period	Estimated requirement assuming constant respiration rate
	Inactive	Active		
A. reticulatus	$48 \cdot 4 \times 18$	$195 \cdot 3 \times 6$	$171 \cdot 2 + 1171 \cdot 8$	$\dfrac{48 \cdot 4 + 195 \cdot 3}{2} \times 24$
	$= 871 \cdot 2$	$= 1171 \cdot 8$	$= 2043 \cdot 0$	$= 2923 \cdot 2$
*A. hortensis**	$17 \cdot 9 \times 18$	100×6	$322 \cdot 2 + 600$	$\dfrac{17 \cdot 9 + 100}{2} \times 24$
	$= 322 \cdot 2$	$= 600$	$= 922 \cdot 2$	$= 1414 \cdot 8$
*M. budapestensis**	$15 \cdot 6 \times 18$	$216 \cdot 8 \times 6$	$280 \cdot 8 + 1300 \cdot 8$	$\dfrac{15 \cdot 6 + 216.8}{2} \times 24$
	$= 280 \cdot 8$	$= 1300 \cdot 8$	$= 1581 \cdot 6$	$= 2788 \cdot 8$

All rates are in μl oxygen/100 mg dry weight at 10°C.

* The figures for the duration of crawling and resting for these species are assumed to be similar to *A. reticulatus*.

The respiration of slugs and snails

The energy available to any animal for tissue growth and reproduction is related to the energy content of the food and the amount of energy used for respiration. Phillipson (1966) has briefly reviewed respiration studies on soil animals.

Most measurements are made for long periods and a value for the oxygen consumed per 24 hours is calculated. Recent work with marine invertebrates (Newell, 1967; Newell and Northcroft, 1967; Halcrow and Boyd, 1967) has shown that the respiration rate as is commonly measured is composed of two components, the 'resting', basal or quiescent rate and the 'active' rate. The duration of these two different rates of oxygen consumption will determine the respiration rate as normally measured by terrestrial ecologists. It is clear that unless the rate of oxygen consumption and the duration of the 'active' and 'quiescent' phases are taken into account the calculated oxygen requirement per 24 hours may be in excess of that actually occurring. I have measured the 'active' and 'quiescent' oxygen consumption of three species of slug with a Gilson differential respirometer and Table 9 shows the estimated oxygen requirement for a 24 hour period. Also in this table is an estimated

TABLE 10. Table showing Q_{10} values for three different species of slug with a dry weight of 20 mg.

Species	'Resting' Q_{10}		'Active' Q_{10}	
	0–10 °C		0–10 °C	
Agriolimax reticulatus	Q_{10}	~ 1	Q_{10}	$= 1 \cdot 7180$
	5–15 °C		5–15 °C	
	Q_{10}	~ 1	Q_{10}	$= 2 \cdot 0160$
	5–15 °C		5–15 °C	
Milax budapestensis	Q_{10}	~ 1	Q_{10}	$= 1 \cdot 7650$
	5–15 °C		5–15 °C	
Arion hortensis	Q_{10}	~ 1	Q_{10}	$= 1 \cdot 7045$

oxygen requirement which is calculated from the mean of the fast and slow respiration rates and is included to show the type of error that can easily occur with conventional respirometry. In all cases this figure is about 1·5 times too high.

A second factor that should be taken into account in the calculation of the metabolic energy requirement is the effect of temperature. As Macfadyen (1963b, p. 76) has pointed out, there are considerable differences in the effects of temperature on invertebrate respiration. These differences may, in part, be due to the differing effects of temperature on the active and quiescent rates of oxygen consumption. Newell (1967) and Halcrow and Boyd (1967) have both shown that the rate of oxygen consumption of a number of invertebrates at rest is relatively unaffected by temperature over the normal environmental range whereas the rate of oxygen consumption of the active organism, like activity itself is strongly temperature dependent. I have found the same relationship in several species of slug and Table 10 shows the $Q_{10}^{0-10\,\circ C}$ and $Q_{10}^{5-15\,\circ C}$ of the 'resting' and 'active' rates of oxygen consumption.

Temperature corrections need be applied therefore, only to the oxygen consumption of the 'active' component of metabolism and since most species only crawl on the surface at night they are subject to relatively small temperature variations (about 10°C). In addition, such crawling occupies only 25% of a 24 hour period so that a temperature correction would add little to the accuracy of the estimate.

References

BAILEY N.T.J. (1952). Improvements in the interpretation of recapture data. *J. Anim. Ecol.* **21**, 120–127.

BARNES H.F. (1953). The absence of slugs in a garden and an experiment in restocking. *Proc. zool. Soc. Lond.* **123**, 49–58.

BARNES H.F. & STOKES B.M. (1951). Marking and breeding *Testacella* slugs. *Ann. appl. Biol.* **38**, 540–545.

BARNES H.F. & WEIL J.W. (1942). Baiting slugs using metaldehyde mixed with various substances. *Ann. appl. Biol.* **29**, 56–68.

BARNES H.F. & WEIL J.W. (1944). Slugs in gardens: their numbers, activities and distribution. I. *J. Anim. Ecol.* **13**, 140–175.

BARNES H.F. & WEIL J.W. (1945). Slugs in gardens: their numbers, activities and distribution. II. *J. Anim. Ecol.* **14**, 71–105.

BERG C.O. (1953). Sciomyzid larvae (Diptera) that feed on snails. *J. Parasit.* **39**, 630–636.

BERG C.O. (1961). Biology of snail killing Sciomyzidae (Diptera) of North America and Europe. *Proc. Int. Congr. Ent. XI (Vienna*, 1960), **I**, 197–202.

BIGOT L. (1963). Observations sur les variations de biomasses des principaux groupes d'invertebres de la 'sansouire' Camargaise. *Terre Vie.* **110**, 319–334.

BIRKETT L. (1957). Flotation technique for sorting grab samples. *J. Cons. perm. int. Explor. Mer.* **22**, 289–292.

BRUEL W.E. VAN DEN & MOENS R. (1958). Remarques sur les facteurs écologiques influençant l'efficacité de la lutte contre les limaces. *Parasitica*, **14**, 135–147.

BURGES A & RAW F. (eds.) (1967). *Soil Biology.* Academic Press, London.

DAHL K. (1917). *Studier og Forsøk øver ørret og ørretvand.* Kristiana.

DEMELOW E.J. (1963). Observations on the feeding of the hedgehog (*Erinaceas europaeus* L.). *Proc. zool. Soc. Lond.* **141**, 291–309.

DOBSON R.M. (1962). Marking techniques and their application to the study of small terrestrial animals. In: *Progress in Soil Zoology.* P.W. Murphy (ed.). Butterworths, London. 228–239.

DOWDSWELL W.H., FISHER R.H. & FORD E.B. (1949). The quantitative study of populations in the Lepidoptera. *Heredity*, **3**, 67–84.

DRIFT J. VAN DER & WITKAMP M. (1960). The significance of the breakdown of oak litter by *Enoicyla pusilla* Barm. *Archs. Néerl. Zool.* **13**, 486–492.

DROZDOWSKI A. (1961). Quantitative studies on the snail fauna of the region of Putawo. *Zesz. nauk. Mikolaja Kopernika Torun.* **6**, 83–148.

DUTHOIT C.M.G. (1961). Assessing the activity of the field slug in cereals. *Pl. Path.* **10**, 165.

FRANÇOIS E., RIGA A. & MOENS R. (1965). Labelling the grey field slug, *Agriolimax reticulatus*, by means of radionuclides. *Parasitica*, **XXI**, 139–151.

FRETTER V. (1952). Experiments with P^{32} and 1^{131} on species of *Helix, Arion* and *Agriolimax*. *Q. Jl. microsc. Sci.* **93**, 133–146.

FRÖMMING E. (1954). *Biologie der mitteleuropaschen landgastropoden.* Duncker & Humblot, Berlin.

FRÖMMING E. (1958). Die Rolle unserer Landschnecken bie der Stoffumwandlung und Humusbildung. *Z. angew. Zool.* **45**, 341–350.

FRÖMMING E. (1962) *Das Verhalten unserer Schnecken zu den Pflanzen ihrer Umgebung.* Duncker & Humblot, Berlin.

GILLIES M.T. (1962). Marking and release experiments with a tropical mosquito by the use of radioisotopes. In: *Radioisotopes in Tropical Medicine.* IAEA, Vienna, 1962.

GOODHART C.B. (1962). Variation in a colony of the snail. *Cepaea nemoralis* L. *J. Anim. Ecol.* **31**, 207–237.

GOULD H.J. (1962). Trials on the control of slugs on arable fields in autumn. *Pl. Path.* **11**, 125.

HALCROW K. & BOYD C.M. (1967). The oxygen consumption and swimming activity of the amphipod *Gammarus oceanicus* at different temperatures. *Comp. Biochem. Physiol.* **23**, 233–242.

HALE W.G. (1967). Collembola, ch. 12 of *Soil Biology*. Burges & Raw (eds.). Academic Press, London.

HAIRSTON N.G., HUBENDICK B., WATSON J.M. & OLIVIER L.J. (1958). An evaluation of techniques used in estimating snail populations. *Bull. Wld. Hlth. Org.* **19** 661–672.

HARVEY L.A. (1964). Natural population of *Cepaea nemoralis*, review and discussion. *Sci. Prog.* Oxford. **52**, 113–122.

HEATH G.W., EDWARDS C.A. & ARNOLD M.K. (1964). Some methods for assessing the activity of soil animals in the breakdown of leaves. *Pedobiologia*, **4**, 80–87.

HERCUS B.H. (1960). Plant cuticle as an aid to determine the diet of grazing animals. *Int. Grassld. Cong.* 1B.

HOWITT H.J. (1961). Chemical control of slugs in orchard grass—Ladino white clover in the Pacific Northwest. *J. Econ. Ent.* **54**, 778–781.

HUNTER P.J. (1966). The distribution and abundance of slugs on an arable plot in Northumberland. *J. Anim. Ecol.* **35**, 543–557.

HUNTER P.J. (1968). Studies on slugs of arable ground. I. Sampling methods. *Malacologia*, **6**, 369–377.

JACKSON C.H.N. (1939). The analysis of an animal population. *J. Anim. Ecol.* **8**, 238–246.

JACKSON C.H.N. (1940). The analysis of a tsetse-fly population. *Ann. Eugen.* **10**, 332–349.

JACKSON C.H.N. (1945). The analysis of a tsetse-fly population. 2. *Ann. Eugen.* **12**, 176–205.

JACKSON C.H.N. (1948). The analysis of a tsetse-fly population. 3. *Ann. Eugen.* **14**, 91–108.

KEVAN D.K.McE. (1962). *Soil Animals*. Wetherby, London.

KNUTSON L.V. (1962). Snail killing sciomyzid flies. *Cornell Plantns*. **17**, 59–63.

KNUTSON L.V., STEPHENSON J.W. & BERG C.O. (1965). Biology of a slug-killing fly, *Tetanocera elata* (Diptera: Sciomyzidae). *Proc. malac. Soc. Lond.* **36**, 213–220.

LESLIE P.H., CHITTY D. & CHITTY H. (1953). The estimation of population parameters from data obtained by means of capture-recapture method. 3. An example of the practical applications of the method. *Biometrika*, **40**, 137–169.

LINCON F.C. (1930). *Calculating water fowl abundance on the basis of banding returns*. (Circ. U.S. Dep. Agric. No. 118.)

LINDEMAN R.L. (1942). The trophic–dynamic aspect of ecology. *Ecology*, **23**, 399–418.

LOMNICKI A. (1964). Some results of experimental introduction of new individuals into a natural population of the Roman Snail, *Helix pomatia* L. *Bull. Acad. pol. Sci. cl. II. Sér. Sci. biol.* **12**, 301–304.

LOŽEK V. (1962). Soil conditions and their influence on terrestrial gastropoda in Central Europe. In: *Progress in Soil Zoology*. P.W. Murphy (ed.). Butterworths, London.

MACFADYEN A. (1963a). The contribution of the microfauna to total soil metabolism. In: *Soil Organisms*. J. Doeksen & J. van der Drift (eds.). North Holland Publishing Co., Amsterdam.

MACFADYEN A. (1963b). *Animal Ecology. Aims and Methods.* 2nd ed. London and New York, 344 pp.

MACFADYEN A. (1964). Energy flow in ecosystems and its exploitation by grazing. In: *Grazing in Terrestrial and Marine Environments. British Ecol. Soc. Symp. IV.* D.J. Crisp (ed.).

MARTIN D.J. (1964). Analysis of sheep diet utilizing plant epidermal fragments in faeces samples. In: *Grazing in Terrestrial and Marine Environments. British Ecol. Soc. Symp. IV.* D.J. Crisp (ed.).

MARTIN A.R. (1961). *The Giant African Snail—a problem in Economic Malacology* Chicago Univ. Press, Chicago.

MILES H.W., WOOD J. & THOMAS I. (1931). On the ecology and control of slugs. *Ann. appl. Biol.* **18**, 370–400.

MILNE A., COGGINS R.E. & LAUGHLIN R. (1958). The determination of numbers of leather-jackets in sample turves. *J. Anim. Ecol.* **27**, 125–145.

MOENS R., FRANÇOIS E., RIGA A. & BRUEL W.E. VAN DEN (1965). Les radioisotopes en écologie animale. Premières informations sur le comportement de *Agriolimax reticulatus. Meded. LandbHoogesch. Opzoekstns Gent.* **30**, 1810–1823.

MOORE C.H. (1934). Slug and beetle. *J. Conch. Lond.* **20**, 85.

MORRIS H.M. (1922). The insect and other invertebrate fauna of arable land at Rothamsted. *Ann. appl. Biol.* **9**, 282–305.

MORRIS H.M. (1927). The insect and other invertebrate fauna of arable land at Rothamsted. Part 2. *Ann. appl. Biol.* **14**, 442–464.

MURPHY P.W. (1962). Extraction methods for soil animals. II. Mechanical methods. In: *Soil Zoology.* P.W. Murphy (ed.). Butterworths, London.

NEWELL P.F. (1965a). *The Behaviour and Distribution of Slugs.* (Ph.D. thesis, London.)

NEWELL P.F. (1965b). Recent methods of marking invertebrate animals for behavioural studies. (Abstract.) *Anim. Behav.* **13**, 579.

NEWELL P.F. (1966). The nocturnal behaviour of slugs. *Med. biol. Illust.* **16**, 146–159.

NEWELL R.C. (1967.) Oxidative activity of poikilotherm mitochondria as a function of temperature. *J. Zool. Lond.* **151**, 299–311.

NEWELL R.C. & NORTHCROFT H.R. (1967). A re-interpretation of the effect of temperature on the metabolism of certain marine invertebrates. *J. Zool. Lond.* 277–298.

NIELSON C.O. (1952–53). Studies on Enchytraeidae. I. A technique for extracting Enchytraeidae from soil samples. *Oikos*, **4**, 187–196.

O'CONNOR F.B. (1967). The enchytraeidae. In: *Soil Biology*, ch. 8. A. Burges & F. Raw (eds.). Academic Press, London.

ODUM H.T. (1957). Trophic structure and productivity of Silver Springs, Florida. *Ecol. Monogr.* **27**, 55–112.

ODUM H.T., CONNELL C.E. & DAVENPORT L.B. (1962). Population energy flow of three primary consumer components of old-field ecosystems. *Ecology*, **43**, 83–96.

ODUM H.T. & SMALLEY A.E. (1959). Comparison of population energy flow of a herbivorous and deposit-feeding invertebrate in a salt marsh ecosystem. *Proc. natn. Acad. Sci. U.S.A.* **45**, 617–622.

OKLAND F. (1929). Methodik einer quontitativen Untersuchung der Landschneckenfauna. *Arch. Molluskenk.* **61**, 121–136.

OKLAND F. (1930). Quantitative Untersuchungen der Landschneckenfauna Norwegens. I. *Z. Morph. Okol. Tiere.* **16**, 748–804.

OWEN G. (1966). Digestion. In: *Physiology of Mollusca*, **2**, ch. 2. K.M. Wilbur & C.M. Yonge (eds.). Academic Press, London.

PAINE R.T. (1965). Natural history, limiting factors and energetics of the opisthobranch, *Narvanax inermis. Ecology*, **46**, 603–619.

PHILLIPSON J. (1966). *Ecological Energetics*. Edw. Arnold Ltd. London, 57 pp.

RAW F. (1955). A flotation extraction process for soil micro-arthropods. In: *Soil Zoology* D.K. McE. Kevan (ed.). Butterworths, London.

ROBERTSON J.D. (1941). The function and metabolism of calcium in the Invertebrata. *Biol. Rev.* **16**, 106–133.

SALT G. & HOLLICK F.S.J. (1944). Studies on wireworm populations. I. A census of wireworms in pasture. *Ann. appl. Biol.* **31**, 52–64.

SATCHELL J.E. (1967). Lumbricidae. In: *Soil Biology*, ch. 9. A. Burges & F. Raw (eds.). Academic Press, London.

SCHNETTER M. (1951). Veränderungen der genetischen Konstitution in natürlichen Populationen der polymorphen Banderschnecken. *Verh. dtsch. Zool. Ges.* 192–206.

SLOBODKIN L.B. (1962). Energy in animal ecology. In: *Advances in Ecological Research. I.* J.B. Cragg (ed.). Academic Press, London.

SOUTH A. (1964). Estimation of slug populations. *Ann. appl. Biol.* **53**, 251–258.

SOUTH A. (1965). Biology and ecology of *Agriolimax reticulatus* (Müll.) and other slugs: spatial distribution. *J. Anim. Ecol.* **34**, 403–417.

SOUTHWARD T.R.E. (1966). *Ecological Methods*. Methuen, London.

TAYLOR J.W. (1894–1921). *Monograph of the Land and Freshwater Mollusca of the British Isles*. **1–3**. Taylor, Leeds.

THOMAS D.C. (1944). Discussion on slugs. 2. Field sampling for slugs. *Ann. appl. Biol.* **31**, 163–164.

TROUGHT T.E.T. & HEATH E.D. (1963). A new approach to the control and assessment of slug damage. *Agricultural Research Council* 12/64.

VÁGVÖLGYI J. (1952). A new sorting method for snails, applicable also for quantitative researches. *Annls. hist.—nat. Mus. natn. hung.* **3**, 101–104.

WALLWORK J.A. (1967). Acari. In: *Soil Biology*, ch. 11. A. Burges & F. Raw (eds.) Academic Press, London.

WEBBE G. & READ W.W. (1962). The use of Co^{60} for labelling snails in the study of field populations. *Ann. trop. Med. Parasit.* **56**, 206–209.

WEBLEY D. (1964). Slug activity in relation to weather. *Ann. appl. Biol.* **53**, 407–414.

WILLIAMSON M.H. (1959). The separation of molluscs from woodland leaf-litter. *J. Anim. Ecol.* **28**, 153–155.

WITKAMP M. (1960). Seasonal fluctuations of the fungus flora in mull and mor of an oak forest. *ITBON. Werkz.* No. 46. Arnhem.

10

A Comparison of Extraction Methods for Terrestrial Arthropods

C.A. EDWARDS & K.E. FLETCHER[1]

I

Introduction

Studies of the productivity of terrestrial communities made as part of the International Biological Programme, will deal with energy flow, nutrient cycling, trophic structure, spatial patterns, relations between species and the diversity of species of plants and animals in chosen ecosystems. For many of these studies the numbers of plants and animals in different communities must be accurately assessed. Plants, the primary producers, can often be counted simply, and although populations of larger animals, the principal secondary producers, are more difficult to estimate, many do not present insuperable difficulties. Populations of small invertebrate animals may contribute a substantial part to the total biomass of the soil ecosystem and to count them is often very difficult, for they are very numerous and their numbers must be estimated by sampling. The number and size of the samples needed varies and depends on the size, activity and distribution of each species. Sampling areas of soil to estimate the numbers of invertebrate animals has been adequately discussed by Hartenstein (1961), Healy (1962), Debauche (1962), Skellam (1962), Hughes (1962) and Macfadyen (1962a). Sampling for particular groups will be dealt with elsewhere in this handbook, as, for example, the detailed treatment of Acari in the chapter by Berthet.

No method yet used, successfully extracts all the invertebrate animals from a soil sample. The immediate aim is to extract as many as possible, and to standardize methods for comparative studies.

This paper describes how the efficiency of all the more common methods of extracting invertebrate animals from soil, in different habitats and soil types, were compared on a scale large enough to assess the influence of the many factors that cause methods to differ in efficiency. From the results the most

[1] This work was made possible by a grant from the Royal Society, England.

suitable methods may be recommended. Methods range from simple sieving or hand-sorting to use of complex pieces of apparatus. The principal methods are of two kinds (Murphy, 1962): those by which the animals are driven from soil, are called dynamic methods; and those by which the animals are physically separated from the soil, i.e. the mechanical methods.

Different stimuli used to encourage the animals to escape from the soil include chemical attractants and repellants, electrical stimuli, vibration and heat, but all the common methods repel the animals from soil samples by changes in moisture content or temperature. The other methods are too specific and too inefficient to merit fuller consideration here.

The earliest funnel method developed by Berlese (1905) consisted of a double-walled funnel containing hot water. The soil or litter rested on a wire mesh and the emerging animals fell into a tube beneath. This funnel was much less efficient than that described by Tullgren (1918) which used heat from a light bulb suspended over the funnel. Most funnels now use Tullgren's principle although with many modifications. For instance, Ford (1937) used a coiled hot wire for the heat source, and Haarlov (1947) showed that a gap between the edge of the sieve and the wall of the funnel prevented water condensing on the walls and trapping the animals after they emerged from the soil. Some of the early Tullgren extractor funnels used ordinary glass laboratory funnels, but Murphy (1950) and others showed that steep-sided funnels were better and Macfadyen (1962) even abandoned the funnel and collected the animals directly into a tube of the same diameter as the sieve that supported the soil. The earlier funnels were large, but there is now a tendency to use many small funnels to allow adequate replication of samples.

There have been many attempts to minimize the amount of soil and debris that falls into the collecting tube (Ulrich, 1933; Valle, 1951; Newell, 1955; Murphy, 1958) but few succeeded enough to be adopted generally, because they also usually lessened the efficiency of extraction. Macfadyen (1953) and Kempson *et al.* (1963) constructed very sophisticated apparatus which completely isolated the upper and lower surfaces of the soil sample and, by cooling and humidifying the lower surface, maintained steep gradients of humidity and temperature in the sample.

The simplest mechanical method of separating small animals from soil is by hand-sorting. This is tedious and inefficient, even when augmented by sieves. Most mechanical methods now rely on differences in density between the animals and the soil, or on the lipophilic properties of cuticles. Impregnated soil samples have also been sectioned and animals counted at the surface, but this method is laborious and unlikely to provide enough data for adequate statistical analysis.

The specific gravity of most small invertebrates lies between 1·0 and 1·1 (Edwards, 1967b) and, as the cuticle is hydrophobic, many animals float from soil stirred in a container of water. When solutions such as brine, sugar or magnesium sulphate, with a specific gravity of about 1·2, are used instead of water, all the animals that are not trapped in the soil float to the surface and can be decanted. Most of the flotation methods in common use are based on this principle. The percentage of animals recovered can be considerably increased by washing the soil through a series of sieves of decreasing mesh size to separate out stones, organic matter and plant material, and break down the soil particles (Morris, 1922). Ladell (1936) further refined the method by agitating the contents of the final container (commonly called a 'Ladell' can) and by bubbling compressed air through the particulated sample immersed in magnesium sulphate solution. This freed the animals from soil and considerably increased the efficiency of the method. Salt and Hollick (1944) described a method and apparatus using these principles that has been widely used with only slight modifications. They added a final stage designed to separate the animals from the organic debris that also floats to the surface of the dense solution, namely shaking the debris and animals with a mixture of xylene or benzene and water. The animals were coated with the organic solvent and, when this rose to the surface, they lay within it just above the water and could easily be picked off from the plant debris, which lay just below the interface in the aqueous phase.

This method is extremely laborious and the efficiency of operators differs. To speed the process, Edwards and Heath (1962) mechanized the washing stage by rotating the Ladell can and sample in a large mesh sieve under a vertical flat fan jet of water. Animals were extracted from the washed soil by shaking soil, debris and animals with a 50/50 mixture of zinc sulphate solution (specific gravity, 1·4) and a mixture of xylene and carbon tetrachloride (specific gravity 1·2), in a specially constructed polythene separating funnel. The animals floated to the surface of the organic solvent and the plant debris to the interface between the zinc sulphate solution and organic mixture; the soil sank to the bottom. The soil and plant debris were then run off leaving the animals in the funnel. Alternatively a siphon system was used to raise the level of the organic mixture until the animals overflowed through a side arm.

Aucamp and Ryke (1964) also used the lipophilic properties of the arthropod cuticle. Small samples of soil were placed in water in perspex containers with sliding plates coated with lanolin, silicone grease or petroleum jelly. After rotating the containers for fifteen minutes the animals stuck to the grease could be examined and counted under a binocular microscope.

Some workers separate invertebrate animals from the soil and debris by centrifuging. Usually air is boiled from the sample which is then centrifuged in a dense solution that supports the animals but not the soil and debris. These methods are suitable only for small soil samples.

Flotation methods are believed to extract a greater proportion of invertebrate animals than dynamic methods, but specimens are often badly damaged and unidentifiable; flotation also collects both dead and live animals and cannot be used for soils with much organic matter. However, flotation methods are much better than dynamic methods for clay soils and allow samples to be stored cold until there is time to deal with them.

Dynamic and flotation methods were compared by Evans (1951), Macfadyen (1953), Edwards (1955), Satchell and Nelson (1962), Wood (1965) and Block (1967), who all compared two methods on a single site or soil type and often had experience of only one of the methods. Such comparisons may be misleading because a method suitable for one type of soil may be unsuitable for another. Flotation methods are unsuitable for enchytraeid worms and nematodes, animals that are only poorly wetted in the oil/water stage.

The efficiency of different extraction methods range from near zero to close to 100% for any particular group of animals, and none is completely successful with all soils and animals. The absolute number of animals in a soil sample is not known, and there is no known way of extracting them all; therefore the efficiency of extraction is difficult to assess. Some workers have put known numbers of animals into soil and then extracted them by different methods, but the animals do not penetrate the soil completely and they are extracted more easily than from soil they were inhabiting naturally.

To compare all the common methods of extraction for different habitats and soil types was beyond our resources and the following studies were made as a compromise. Many soil ecologists were asked for details of their methods and on the basis of their replies, eleven commonly used extraction methods were tested with soil from woodland, pasture and arable land all on the same soil type. The efficiency of extraction of the commonest dynamic and flotation methods was compared using five different soil types ranging from sands and peats to clays. Factors that influenced the efficiency of a standardized form of the Tullgren apparatus were also studied.

The information which showed the best methods for each group of animals in particular habitats or soil types is given in the last section of this paper. The complexity of the different methods, their availability and practicability are also discussed. Definite advice is given about methods for use in the PT section of the International Biological Programme.

II

Analysis of questionnaires

Of 150 questionnaires sent out to soil zoologists and ecologists, 82 were completed and returned. Questions were asked about general methods of extraction, about those used to study particular groups and about details of apparatus and methods and any major modifications. For example, questions were asked about the size and shape of Tullgren funnels, the power and distance from the sieve of the source of heat, whether the funnels were cooled, the time taken for complete extraction, and the preservative liquid in the collection vials. Details of flotation methods were asked for, such as the nature and specific gravity of the dense solution and whether an oil/water stage was used.

Replies were sometimes incomplete, sometimes more information was given than was requested, but we thank all who answered. The frequency with which different methods are used is summarized in Fig. 1. Most workers used a single method; 74% used some form of Tullgren funnel and 61% of these used a relatively standard kind, usually with electric light as a heat source. The few that used neither light nor heat were mostly concerned with taxonomy. Seven workers used modified Macfadyen funnels with steep gradients of heat and humidity produced by cooling and humidifying the completely separate lower part of the sample.

Four workers used simple flotation with dense solutions for general studies, but several others used flotation with sugar solution for Collembola and other fragile animals which were recovered in good condition for taxonomic studies. Except for dynamic methods Salt and Hollick flotation was by far the most popular method (12%). Several workers hand-sorted the soil for the larger animals, either in the field or in the laboratory. Most of the other methods were used only by those workers who devised them; for example, the Muller and Naglitsch centrifugal flotation method, d'Aguilar flotation, and grease film extraction.

METHODS OF EXTRACTION
MOST COMMONLY USED

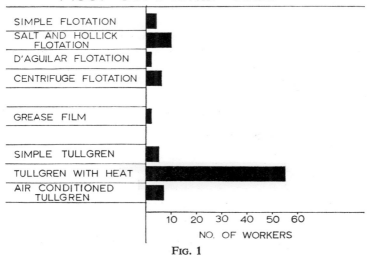

FIG. 1

SIZE OF TULLGREN FUNNELS

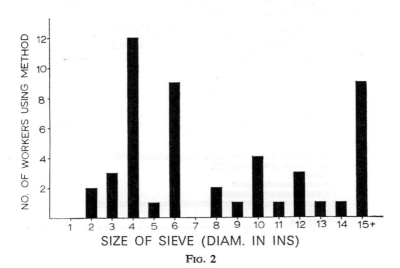

FIG. 2

Chapter 10

The Tullgren apparatus and the method of using it varied greatly (Fig. 2). Funnels ranged from 5 cm to more than 37·5 cm in diameter, but 10 cm was the most common diameter and 15 cm diameter funnels were often used. Small funnels and more samples are becoming more popular.

Workers were asked whether soil samples were placed in the funnels with as little disturbance as possible or were broken up, and of seventeen who answered, nine did not disturb the samples (Fig. 3). The power of heat source varied (Fig. 4); most workers used electric light bulbs of less than 35 watts, but some used 100 watts and one 200 watts, so the time taken to complete the extraction (Fig. 5) consequently ranged from 1 day to more than 12 days.

METHOD OF PLACING SAMPLE IN SIEVE

Fɪɢ. 3

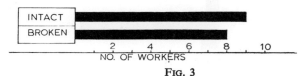
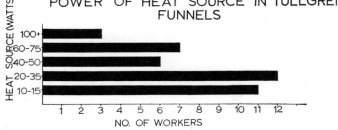

Fɪɢ. 4

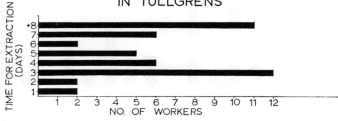

Fɪɢ. 5

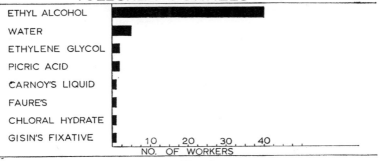

FIG. 6

Most Tullgren funnels took either 3–4 days to extract all the animals or more than seven. Many workers taking longer to extract their samples claimed a more efficient extraction. The replies suggest that many workers extracted for an arbitrary time without fully testing whether it was adequate for all soils they investigated.

Most workers used ethyl alcohol in the collecting vials under the funnels, a few used water and a minority, other fixatives (Fig. 6).

The questionnaires enabled us to choose the methods to test. In general, dynamic methods were preferred, probably partly because they were easy to operate and because specimens were undamaged. Our studies showed that funnels are comparatively efficient, and because most participants in the International Biological Programme will use them, factors affecting their efficiency were studied.

III

Description of extraction methods tested

(1) Dynamic methods

These are all based on the original Tullgren principle but differ enough to justify testing five versions in several ways.

(A) *The simple plastic funnel extractor*

Sixteen domestic plastic funnels are mounted upon two plywood boards set one above the other with holes to support each funnel. The funnels, $16 \cdot 5 \times 19 \cdot 8$

cm deep, have an included angle of 60°, and narrow to an opening 1·2 cm diam., where the collecting vessel, 2·5 cm diam. × 5 cm deep is held in position by a rubber bung. The soil samples are contained in anodized aluminium sample containers, 12·8 cm diam. × 7·5 cm deep, and rest upon a 10 mesh, 23 S.W.G. stainless steel sieve. The sieves are held in position by wire supports from three outwardly directed lugs 1·2 cm long, at the base of the sample container. The samples were extracted for nine days, at a room temperature thermostatically maintained at about 20°C and the animals were collected into vials containing 70% v. aqueous ethyl alcohol and glycerol (in the ratio 20: 1).

(B) *The Rothamsted funnel extractor without light or heat source*

This battery of sixteen anodized aluminium funnels is mounted upon two plywood boards, set in an angle-iron framework, one above the other, with holes 12·8 cm diam, and 4·2 cm diam., respectively, to support each funnel. The funnels 15 cm diam. × 25 cm deep, with an included angle of 32°, narrow to an opening 1·2 cm diam., where a collecting vial, 2·5 cm diam. × 5 cm deep, is held in position by a rubber bung. The soil samples rest on a 10 mesh S.W.G. stainless steel sieve in anodized aluminium sample containers, 12·8 cm diam. × 7·5 cm deep. The sieves are held in position by wire supports from four outwardly directed lugs, 2·5 cm long, at the base of the sample container. The sample container is supported in the top of the funnel by the lugs, leaving a gap between the sieve and funnel that minimizes condensation. The samples were extracted for nine days at a room temperature thermostatically maintained at 20°±1°C. The animals were collected into vials containing 70% v. aqueous ethyl alcohol and glycerol (in the ratio 20: 1).

(C) *The Rothamsted funnel extractor with light and heat source*

The apparatus is similar to that in B above but over each sample a tinned cylinder 26·6 cm high × 12·8 cm diam. is suspended, containing a 25 w. pearl electric light bulb 25 cm above the surface of the soil. The light bulbs, wired in parallel from a 240 v. mains were controlled by a rheostat. The extraction time was 5 days. The temperature of the soil nearest the bulb increased from about 35°C after the first 12 hours to about 39°C after 5 days. The room temperature was controlled at 20°±1°C during the whole of the extraction period.

(D) *The Murphy split funnel extractor* (Murphy, 1958)

This apparatus has 18 small funnels mounted on a wood and metal framework, and a similar battery of 18 light cylinders suspended above. The cylinder unit is counterpoised and can be raised for access to the funnels and lowered to cover the sample containers. The funnels, 8 cm diam. × 24·2 cm, narrow to an opening 1·6 cm diam., where a collecting tube, 2·5 cm diam. × 5 cm deep, is held in position by a rubber bung. The sample container, 5 cm diam. × 5 cm deep, has three brackets at its base supporting a sieve plate, 14 mesh, 26 S.W.G., held 0·5 cm below the lower rim of the sample container leaving a gap at the sides. This unit is supported on a large sieve plate, 4 mesh, 20 S.W.G., which rests at the base of the upper portion of the funnel. The cylinders 7·5 cm diam. × 30 cm deep each contain a 15 w. electric light bulb. The soil samples, 5 cm diam. × 5 cm deep, were extracted intact and inverted with vials containing 70% v. ethyl alcohol and glycerol (in the ratio 20: 1) below the funnels. The extraction took six days in a room at a temperature of 16–20°C with the air temperature at the top of the sample about 30°–35°C.

(E) *The Kempson infra-red extractor* (Kempson *et al.*, 1963)

The apparatus uses a similar principle to the Macfadyen canister but is much larger. It consists of a rectangular cabinet of fibre board open at the top, supported by a framework of slotted angle. A fibre baffle board separates the cabinet into two compartments. In the lower compartment a fibreglass tray, 90 cm × 54 cm, is used as a continuous-flow cold water bath. In the upper compartment two 250 w. infra-red lamps, are suspended 51·5 cm above the baffle board and these are wired to a 'Sunvic Simmerstat' time switch (type TXY–B) and to two 100 w. light bulbs, so that when the infra-red lamps are off, the 100 w. bulbs are on. The sample holders are round plastic dishes, 12·5 cm diam. × 5 cm deep, with tapered sides. The bottom of each dish, cut away and replaced by a grid of stiff plastic with approximately 1 cm mesh, fits into holes carefully cut in the fibre baffle board. Below each sample container there is another plastic bowl joined to it by a tightly fitting rubber ring and containing saturated picric acid solution. Access to the sample holders is through a vertically sliding door in the front of the cabinet. The bowls containing the collecting fluid are partly immersed in the cold water bath and the soil samples are placed upon a circle of cotton fillet net (11 threads per 2·5 cm each warp and weft) supported by the plastic grid. The gap between the sides of each soil sample and the sides of the container is filled with expanded polystyrene with holes cut to fit the sample to reduce heat transfer between the upper and lower compartments. On the 3rd day of extraction the simmer-stat was switched on at a low setting. During the fourth and fifth days the heating was increased and by the end of the sixth day the lamps were left on all the time until the end of the extraction on the eighth day. Temperatures at the soil surface were about 70°C at the end of the extraction period.

(F) *The Macfadyen high gradient canister extractor*

The apparatus, based on Macfadyen's (1962b) high gradient canister extractor as described by Block (1966), does not use funnels. Thirty-two extractor units arranged in four rows of eight, each contain a cylindrical soil core 3·8 cm diam. × 3·0 cm deep. The units are held in two trays which form the water baths for cooling the collecting canisters. A horizontal asbestos baffle separates the heating from the cooling compartment. The intact soil cores are inverted over a metal gauze (25 mesh per cm²) on the top of an aluminium canister 4·0 cm diam. × 8 cm deep containing the collecting fluid. The sample holder of heat-resistant laminated phenolic plastic 'Tufnol', is fixed to the top of the canister by a watertight rubber sleeve. The units are held in position by holes in the asbestos baffle. A steep temperature gradient is maintained throughout the soil sample by heating from above by a 60 w. pearl electric light bulb, and cooling the collecting canisters below by circulating cold-water through the bath. A 'Zenith' variac transformer (type 100M) controls the voltage passing through the electric light bulb above each unit. The animals drop into distilled water. Samples were extracted for three days, with heating controlled at the following voltages: 0–24 hours at 60 volts, 24–48 hours at 100 volts and 48–72 hours at 140 volts. The temperature at the sample surface was approximately 120°C at the end of the extraction period.

Temperature changes
Sample

Time	Top	Bottom	Gradient
24 hrs	32°C	20°C	12°C
48 hrs	70°C	30°C	40°C
72 hrs	120°C	50°C	70°C

(G) *The Macfadyen air-conditioned funnel extractor* (Macfadyen, 1962b)

The apparatus is completely enclosed in a rectangular cabinet, made of chip-board, supported on an angle-iron framework, and lined with expanded polystyrene. The funnels arranged in rows of three, fit tightly into holes in removable chip-board trays and are 25 cm high × 15 cm diam. with an included angle of 32° narrowing to an opening 1·2 cm diam. Collecting tubes 2·5 cm diam. × 5 cm deep are held in position at the bottom of the funnels by rubber sleeves. A door in the front gives access to the funnels in the front and the sample containers, 10·5 cm diam. × 5 cm deep, are suspended above funnels, and mounted in a baffle board covered with aluminium foil to minimize heat transfer. The sample containers are supported by a metal rod across the top of each funnel and held firmly in position by the holes in the baffle. The upper part of the cabinet containing the heater assembly is supported on pulleys and can be raised to insert samples and lowered onto the sample containers, which fit flush to the surface of the baffle board. The heaters are lengths of 20 gauge nichrome wires 7·5 ohm stretched between ceramic insulators. Each heater serves three funnels and are arranged in pairs each side of vertical ducts. A fan circulates moist air in the lower compartment with ducts arranged so the air currents pass over a cold water bath, through the coils of an evaporator from a refrigerator, to the lower surface of the soil sample. The air then passes through holes in a false back plate and back to the fan. The air passages in the false back are of equal length, to ensure an even air flow to all funnels. The evaporator coils lead to long lengths of flexible copper pipe to complete the circuit between the compressor and condenser. The compressor is driven by a $\frac{1}{4}$ h.p. capacitor start induction motor (230 v. 50 cycles single phase) and a thermostat controls the air temperature in the cabinet.

The soil samples were taken in the field in the sample containers, which had a sieve of 10 mesh, 23 S.W.G., held in position by cross-wire supports from the sides. Intact samples were inverted and left in the apparatus for nine days and animals were collected in 70% v. aqueous ethyl alcohol and glycerol (ratio 20: 1). For the first seven days the air in the lower cabinet was regularly moistened, using an atomizer driven by a compressor to maintain a relative humidity greater than 85% and cooled (8–10°C) while the temperature in the heating compartment increased to 45°C. The refrigerator unit was switched off on the seventh day and the relative humidity allowed to fall.

(2) Mechanical methods

(H) *Simple brine method of flotation*

The soil sample, immersed in a saturated sodium chloride solution (S.G. 1·2) in a galvanized can is stirred until most of the soil is in suspension in the solution; it is then allowed to settle, and the floating material decanted into a 150 mesh sieve container. The process is repeated to ensure all floating material is collected. The collected material is washed free from brine solution and carefully transferred, together with some water, to a 500 ml conical flask. Xylene is added and the mixture shaken. The contents of the flask are poured into a 1 litre beaker and the xylene-water interface brought to the top of the beaker by adding more water. The interface is examined under a binocular microscope and the animals picked out with a small wire spoon.

(I) *The Salt and Hollick method of flotation*

The apparatus is modified slightly from that described by Salt and Hollick (1944). The soil samples are put in cans and water added. They are then frozen at $-10°C$ to break up the soil particles and kept at this temperature. Before flotation each soil sample is thawed and the soil washed under high-pressure jets of water through two sieves with successively smaller mesh into a reservoir below. The meshes of the two sieves are 7 mm and 2·5 mm respectively. Material retained by the 7 mm mesh is washed, teased apart if necessary and discarded and fragments retained by the 2·5 mm mesh are washed into the reservoir. The excess water and suspended material overflow into another deep container with an overflow lip and a 120 mesh sieve in the bottom (43 S.W.G.) ('Ladell' container). The water and fine silt particles pass through the 120 mesh sieve and run out to a drain. After all material is washed from the sieves the first container is tilted and the contents washed into the 'Ladell' container, which is then drained and placed on a pivoted stand. A saturated solution of magnesium sulphate (S.G. 1·2) is used to float the organic material to the surface. Compressed air, passed through the base of the 'Ladell' container, agitates the soil and free organic particles.

The material floating to the surface is decanted into a cylindrical glass tube, diam. 5 cm × height 17·5 cm, covered at the base with a piece of 180 mesh bolting silk. The organic matter collected is washed and shaken in a flask with water and xylene. The animals are picked off the xylene-water interface under a binocular with a small wire spoon.

(J) *The Edwards and Heath method of flotation*

This apparatus (Edwards and Heath, 1962) is a mechanized Salt and Hollick apparatus. The 'Ladell' container is of black polythene with a 180 mesh sieve at the bottom. Four of these modified 'Ladell' containers rest in rotating funnels driven by a common shaft geared to a $\frac{1}{4}$ h.p. electric motor so they rotate on the vertical axis at 20 r.p.m. The soil samples, stored at $-10°C$, are thawed and placed in wire mesh (4 mesh, 23 S.W.G.) baskets standing on the 'Ladell' sieve. Each soil sample is sprayed with water from two fixed flat fish-tail jets. One plays across the base sieve to prevent it blocking with silt and the other is directed vertically downwards to break up the soil as it rotates. The water and fine silt particles pass through the sieve, along channel-ways to a settling tank, and then to the drain. The water jets are fixed to a counterpoised bar, which can be raised to allow the 'Ladell' containers to be removed and samples replaced. Each soil sample was washed for 10–30 min. according to soil type, and any material retained in the wire basket was washed above the 'Ladell' container, teased apart, if necessary, and discarded. The subsequent procedure was similar to the Salt and Hollick method, with a flotation, and oil-separation stage.

(K) *The grease film method of extraction* (Aucamp and Ryke, 1964)

This apparatus consists of two rectangular plastic tanks (7·5 cm × 10 cm × 20 cm) each containing two removable plastic plates 20 cm × 10 cm which slide into tracks in the tank so they are held against the inner surfaces. These plates are coated on the inner surface with a thin layer of silicone or other grease such as Shell white petroleum jelly. The soil samples (5 cm × 5 cm), stored at $-10°C$, are thawed, and allowed to stand in beakers of water, with occasional stirring, for 1 hour. This breaks up the soil particles and separates any tangled root present. The water-soil mixtures are poured into the two plastic tanks and water added to make them of equal weight. A lid, fitted onto each tank, was kept watertight by a rubber seal and clamp. The tanks are then fitted into a frame rotating about its horizontal axis and which can be tilted to a vertical position to facilitate loading. The frame rotates on a central axis, driven by an electric motor at 16 revolutions per minute and samples are rotated for 15 minutes. The grease plates are removed and examined under a binocular microscope and most animals can be identified while still embedded in the grease.

IV

A comparison of all methods on a single soil type

The methods described in Section III were compared in woodland, pasture and arable habitats all on a silt-clay loam soil. From each site, samples were taken with the appropriate coring tools as close to one another as possible, from a small uniform area, and were immediately placed in aluminium tins and taken to the laboratory. Randomly selected batches of sixteen were placed in each of the sets of funnels; the remaining samples, for extraction by flotation and grease-film, were sorted into similar batches and stored at $-10°C$. Sixteen samples are the minimum number required to show a difference in efficiency of 20% to be significant between methods for most species of the micro- and meso-fauna: more are needed for sparse or aggregated animals. Samples were taken only to a depth of two inches because more than 80% of the arthropods were in this layer.

KEY TO FIGURES 7-16

DYNAMIC: A. Plastic Funnels   B. Tullgrens (no heat source)   C. Rothamsted Tullgrens   D. Murphy Split Funnels   E. Kempson Infra-red   F. Macfadyen Canisters   G. Macfadyen Funnels.

MECHANICAL: H. Brine Flotation   I. Salt & Hollick Flotation   J. Heath & Edwards Flotation   K. Aucamp Grease Film.

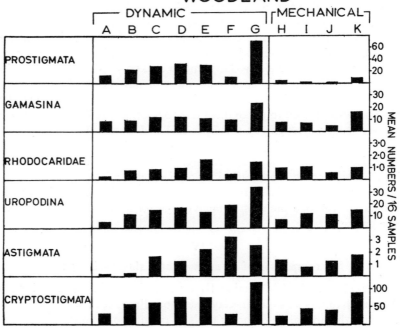

FIG. 7

COMPARISON OF METHODS
WOODLAND

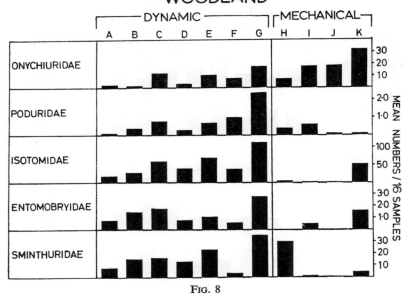

FIG. 8

COMPARISON OF METHODS
WOODLAND

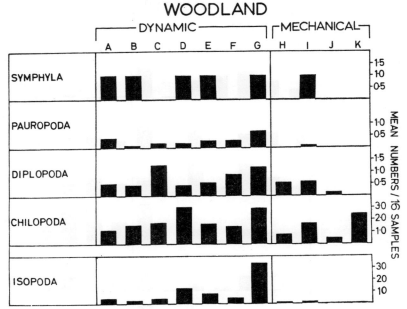

FIG. 9

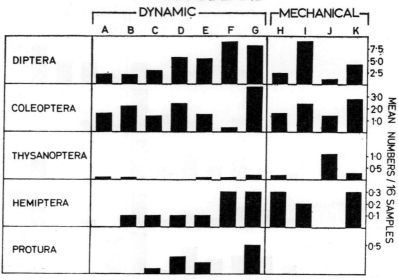

COMPARISON OF METHODS
WOODLAND

Fig. 10

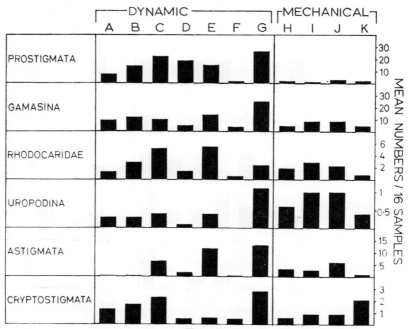

COMPARISON OF METHODS.
PASTURE.

Fig. 11

COMPARISON OF METHODS.
PASTURE.

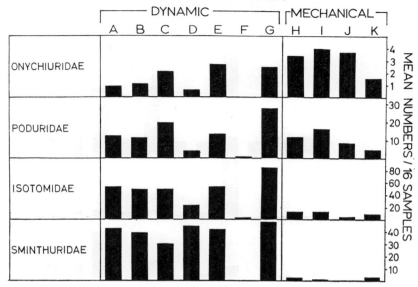

FIG. 12

COMPARISON OF METHODS.
PASTURE.

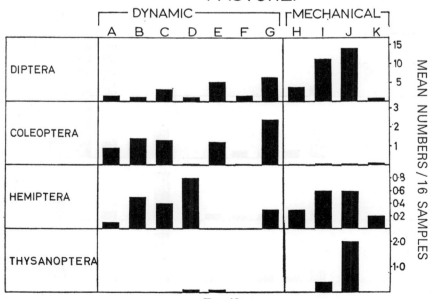

FIG. 13

COMPARISON OF METHODS.
FALLOW.

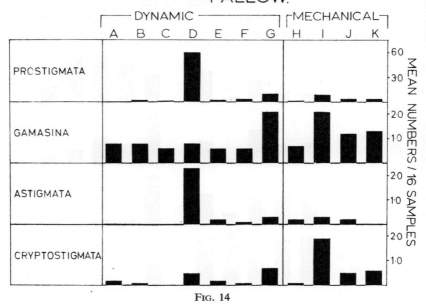

FIG. 14

COMPARISON OF METHODS.
FALLOW.

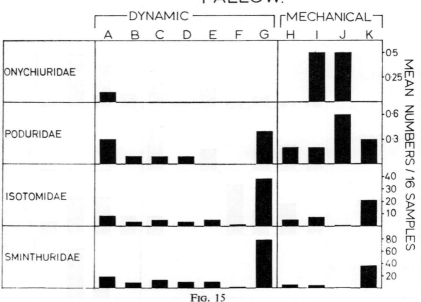

FIG. 15

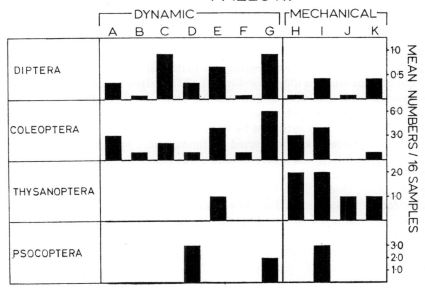

COMPARISON OF METHODS.
FALLOW.

FIG. 16

The animals, extracted from the samples by different methods, were stored in 70% aqueous ethyl alcohol with 5% glycerol, before being sorted into taxonomic classes. In this stage of the investigation 576 samples each containing 50—1000 animals were extracted and the animals identified as shown in Figs. 7–16. The numbers of animals from the small samples were multiplied by volume conversion factors to allow methods to be compared.

Animals are not usually randomly distributed in soil and there were large numbers of some groups in a few samples. All the data were transformed to $\log (n+1)$, and an analysis of variance for the factors, site, method and species revealed highly significant differences between sites, methods and species, and also for the interactions. Woodland soil had most animals, and the arable fewest. No method extracted all animals and absolute efficiency is not known, but relative efficiency for different groups of animals was tested with Tukey's (1953) w-procedure, which involves the calculation of an 'honestly significant difference' (hsd) based on the standard error of a treatment mean multiplied by a factor obtained from a table of the 'Upper Percentage Points of the Studentized Range'. The comparisons are summarized for the three habitats in Tables I, II and III.

Mechanical methods, thought to be the more efficient on heavy clay soils, were less efficient for these soils than the best dynamic methods; but they were as efficient as, or more so than, the dynamic methods, only on the fallow site probably because there was less organic matter to interfere with the flotation or to become stuck in the grease-film. The Macfadyen air-conditioned funnels extracted more Collembola and Acarina from both woodland and pasture soils than any of the other methods, except that the onychiurid Collembola were more efficiently extracted by flotation methods.

More myriapods were extracted by funnels than by mechanical methods, except from fallow soil where flotation was better for millipedes and centipedes. Flotation methods were generally better for the larger arthropods, particularly Insecta but both flotation and funnel methods were fairly efficient for these groups. The most efficient method for all groups of animals was the Macfadyen air-conditioned funnel apparatus but the grease-film method was promising for microarthropods although they were not recovered in good condition for taxonomic study and only small soil samples could be used.

V

A comparison of flotation and dynamic methods on different soil types

In Section IV eleven methods were compared with soil from three habitats on a single soil type. The efficiency of a particular method differed with soil, with different amounts of organic matter, and under different types of plant cover. This differing efficiency of extraction with soil of different types was also tested with the Rothamsted pattern Tullgren funnels and Salt and Hollick flotation method, representative of dynamic and mechanical methods. Sixteen pairs of adjacent soil samples 10 cm diam. and 5 cm deep were taken at random from small uniform areas on each of five sites all with different soils, and where possible from more than one habitat on each soil type. One sample from each pair was extracted (intact and inverted) on the Rothamsted pattern funnels with heat applied; the other was frozen at $-10°$ C and the animals later extracted by the Salt and Hollick technique described previously.

The results, summarized in Table IV, agree closely in many respects with those obtained when eleven methods were compared in Section IV. For all

TABLE I. Comparison of methods—woodland, silt clay loam soil

Animal Group	Simple Plastic Funnel	Rothamsted Funnels (No Heat)	Rothamsted Funnels (Heat)	Murphy Split Funnels	Kempson Infra-red Extractor	Macfadyen High Gradient Canister	Macfadyen Air Conditioned Funnels	Simple Brine Flotation	Salt and Hollick Flotation	Edwards and Heath Flotation	Grease Film Extractor	Significance of F
Isopoda	*	*	*	—	*	*	H	*	*	*	*	0·01
Symphyla	—	—	—	—	H	—	—	—	—	—	—	—
Pauropoda	—	—	—	—	—	—	H	*	*	*	*	0·05
Diplopoda	—	—	H	—	—	—	—	—	—	*	*	0·01
Chilopoda	—	—	—	H	—	—	—	—	—	*	—	0·05
Prostigmata	*	*	—	—	*	*	H	*	*	*	*	0·01
Mesostigmata (Gamasina)	*	*	—	—	—	—	H	*	*	*	—	0·01
Mesostigmata (Uropodina)	*	*	—	—	*	—	H	*	*	*	—	0·01
Astigmata	*	*	—	—	—	H	—	—	*	—	—	0·01
Cryptostigmata	*	*	*	—	*	*	H	*	*	*	—	0·01
Onychiuridae	*	*	—	*	—	*	—	*	—	—	H	0·01
Poduridae	*	*	—	*	*	—	H	*	*	*	*	0·01
Isotomidae	*	*	—	*	—	*	H	*	*	*	—	0·01
Entomobryidae	*	—	—	*	—	*	H	*	*	*	—	0·01
Sminthuridae	*	—	—	*	—	*	H	—	*	*	*	0·01
Protura	*	*	—	—	—	*	H	*	*	*	*	0·01
Psocoptera	—	—	—	—	—	—	—	—	—	—	—	
Hemiptera	—	—	—	—	—	H	—	—	—	—	—	—
Coleoptera	—	—	—	—	—	*	H	—	—	—	—	0·01
Diptera	*	*	—	—	—	—	—	*	H	—	—	0·01
Thysanoptera	*	*	*	*	*	*	*	*	*	H	*	0·01

H = Greatest numbers recovered, on average.
— = Not significantly less than H at 5% probability level.
* = Significantly less than H at 5% probability level.

TABLE II. Comparison of methods—pasture, silt clay loam soil.

Animal Group	Simple Plastic Funnel	Rothamsted Funnels (No Heat)	Rothamsted Funnels (Heat)	Murphy Split Funnels	Kempson Infra-red Extractor	Macfadyen High Gradient Canister	Macfadyen Air Conditioned Funnels	Simple Brine Flotation	Salt and Hollick Flotation	Edwards and Heath Flotation	Grease Film Extractor	Significance of F
Isopoda	—	—	—	—	—	—	H	—	—	—	—	—
Pauropoda	—	H	—	—	—	—	—	—	—	—	—	—
Prostigmata	*	—	—	—	—	*	H	*	*	*	*	0·01
Mesostigmata (Gamasina)	—	—	—	*	—	*	H	*	—	—	*	0·01
Mesostigmata (Uropodina)	—	—	—	—	—	*	H	—	—	—	—	
Astigmata	*	*	—	*	—	*	H	*	*	—	*	0·01
Cryptostigmata	—	—	—	*	—	*	H	—	—	—	—	0·05
Onychiuridae	—	—	—	*	—	*	—	—	H	—	—	0·01
Poduridae	—	—	—	*	—	*	H	—	—	*	*	0·01
Isotomidae	—	—	—	*	—	*	H	*	*	*	*	0·01
Entomobryidae	—	—	—	—	—	—	H	—	H	—	—	—
Sminthuridae	—	—	—	—	—	*	H	*	*	*	*	0·01
Protura	—	—	—	—	—	—	—	—	H	—	—	—
Psocoptera	—	—	—	—	—	—	—	—	—	H	—	—
Hemiptera	—	—	—	H	*	*	—	—	—	—	—	0·05
Coleoptera	*	—	—	*	—	*	—	*	*	*	H	0·01
Diptera	*	*	*	*	—	*	—	*	—	H	*	0·01
Thysanoptera	*	*	*	*	*	*	*	*	—	H	*	0·01

H = Greatest numbers recovered, on average.
— = Not significantly less than H at 5% probability level.
* = Significantly less than H at 5% probability level.

TABLE III. Comparison of methods—fallow, silt clay loam soil

Animal Group	Simple Plastic Funnel	Rothamsted Funnels (No Heat)	Rothamsted Funnels (Heat)	Murphy Split Funnels	Kempson Infra-red Extractor	Macfadyen High Gradient Canister	Macfadyen Air Conditioned Funnels	Simple Brine Flotation	Salt and Hollick Flotation	Edwards and Heath Flotation	Grease Film Extractor	Significance of F
Symphyla	—	—	—	—	H	—	—	—	—	—	—	—
Pauropoda	—	—	H	—	—	—	—	—	—	—	—	—.
Diplopoda	—	—	—	—	H	—	—	H	—	—	—	—
Chilopoda	—	—	—	—	—	—	—	—	H	—	—	—
Prostigmata	*	*	*	H	*	*	*	*	*	*	*	0·01
Mesostigmata (Gamasina)	—	—	—	—	—	—	—	—	H	—	—	0·05
Mesostigmata (Uropodina)	—	—	—	—	—	—	—	—	H	—	—	—
Astigmata	*	*	*	H	*	*	*	*	*	*	*	0·01
Cryptostigmata	*	*	*	*	*	*	—	*	H	*	*	0·01
Onychiuridae	*	*	*	*	*	*	*	*	H	H	*	0·01
Poduridae	—	—	—	—	*	*	—	—	—	H	—	0·05
Isotomidae	*	*	*	*	*	*	H	*	*	*	—	0·01
Entomobryidae	—	—	H	—	—	—	—	—	—	—	—	—
Sminthuridae	*	*	*	*	*	*	H	*	*	*	—	0·01
Psocoptera	—	—	—	H	—	—	—	—	—	—	—	—
Hemiptera	—	—	—	—	—	—	—	—	—	—	H	—
Coleoptera	—	—	—	—	—	—	H	—	—	—	—	—
Diptera	—	—	—	—	—	—	H	—	—	—	—	0·01
Thysanoptera	—	—	—	—	—	—	—	—	H	—	—	—

H = Greatest numbers recovered on average.
— = Not significantly less than H at 5% probability level.
* = Significantly less than H at 5% probability level.

TABLE IV. Comparison of methods on different soil types.

Soil Type:	Peat			Clay Mull Over Limestone	Clay Loam		Silt Clay Loam			Sand Loam		
Habitat:	Woodland	Pasture	Fallow	Woodland	Pasture	Fallow	Woodland	Pasture	Fallow	Woodland	Pasture	Fallow
Animal Group												
Symphyla	0	0	0	—	—	—	0	0	0	0	0	0
Pauropoda	0	0	0	—	—	—	—	0	0	0	—	0
Diplopoda	0	0	0	—	—	—	—	0	0	0	0	0
Chilopoda	(F)**	0	—	—	—	—	—	0	0	—	0	—
Prostigmata	(T)**	(T)**	(T)**	(T)**	(T)**	(F)*	(T)**	(T)**	—	(T)**	(T)**	(T)**
Mesostigmata (Gamasina)	(F)**	(T)**	—	—	—	—	(T)**	—	—	—	(T)**	(T)**
Mesostigmata (Uropodina)	(F)**	—	0	—	0	—	—	—	0	—	—	0
Astigmata	(F)**	(T)*	(T)**	—	0	(F)**	—	—	—	—	(T)**	(T)**
Cryptostigmata	(F)*	—	(T)**	—	—	(F)*	(T)**	—	(F)**	(T)*	(T)*	—
Onychiuridae	(F)**	—	(F)**	—	(F)*	(F)**	—	—	—	—	—	(T)**
Poduridae	(F)*	(F)*	0	(T)*	(F)*	—	—	—	—	(T)**	—	(T)**
Isotomidae	(T)**	(T)**	(T)**	—	—	—	(T)**	(T)**	—	—	(T)**	(T)**
Entomobryidae	(T)*	—	0	—	(F)*	—	0	0	0	—	(T)*	0
Sminthuridae	0	(T)**	(T)**	—	(T)**	—	(T)**	(T)**	(T)*	0	(T)**	—
Hemiptera	—	—	(F)**	(F)*	(F)**	0	0	0	0	—	—	0
Thysanoptera	—	(T)**	—	(F)**	—	—	0	0	—	(F)*	(T)**	(T)*
Diptera	(T)**	—	—	(F)**	—	—	—	(F)**	—	(F)**	(F)*	—
Coleoptera	—	(T)**	(T)**	—	(F)**	(F)*	—	—	—	(F)*	—	—
Protura	0	0	0	—	0	—	0	0	0	—	0	0

— = No significant difference between Means at 5% and 1% levels.
* = Significant difference between Means at 5% level.
** = Significant difference between Means at 1% level.
(T) = Tullgren significantly more efficient.
(F) = Flotation significantly more efficient.
0 = Not known.

the mineral soils the flotation method usually extracted more insects and their larvae but from peat soil, Tullgren methods extracted more. The efficiency of flotation and funnel methods did not differ significantly for many soils and insect groups.

More Collembola and Acarina were extracted by the funnel method from most soils, but with fallow soils the flotation method was more efficient, probably because there is less organic matter in such soil. In contrast to other microarthropods, the less active onychiurid and podurid Collembola tended to be extracted more efficiently by flotation, probably because they are trapped in the soil during funnel extraction. In general, funnel methods are preferable for studying populations of micro-arthropods except with soils containing little organic matter, and flotation methods are preferable for Insecta and larger arthropods, except in soils with much organic matter.

VI

Storage of soil samples

The animals should be extracted from soil soon after the samples are taken but often this may be impossible. The animals can be satisfactorily extracted mechanically even after storage at $-10°C$ for several months. When storing samples for extraction in funnels all the animals must be kept alive, but their numbers change as eggs hatch and individuals die. Ideally all samples should be extracted the day they were taken but if this is impossible changes that occur during storage must be known. Consequently these were assessed.

Many cores, 5 cm diameter, 15 cm deep, were taken very close together from pasture on a clay loam soil. They were mixed randomly and divided into four equal batches and kept at 5, 10, 15 and 20°C. Sixteen samples were also extracted in Rothamsted pattern funnels immediately after they were brought in from the field and a set of sixteen stored samples, from each temperature, extracted in funnels each week thereafter. Results are now complete only for the control samples and those stored at 5° and 20°C for 28 and 56 days; Figs. 17 and 18 summarize the results. After storage for 28 days at 5°C only rhodacarid mites showed a significant ($P = 0·05$) decrease in numbers. By 56 days numbers of gamasid mites and Diptera larvae had increased significantly, the larvae presumably from hatched eggs already in the soil. Numbers of Thysanoptera had decreased significantly after 56 days. There were slightly fewer Collembola after storage probably because predatory gamasid mites increased (Edwards, 1967a).

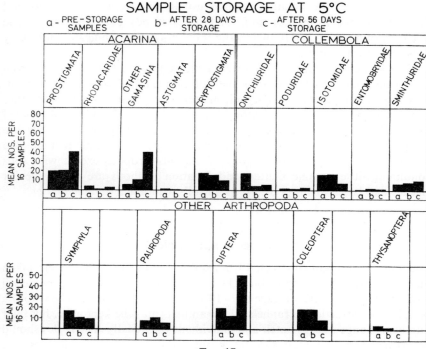

FIG. 17

In samples stored at 20°C, numbers of animals changed much more. Gamasid and rhodacarid mites had increased very rapidly after 28 days and by 56 days there were significantly more astigmatid mites. Numbers of Sminthuridae had lessened significantly by 28 days and those of Symphyla by 56 days. Numbers of prostigmatid mites and enchytraeid worms also increased but not significantly. These results show that storage for up to a week at temperatures of about 5°C should not cause any serious changes in numbers of animals in the samples.

VII

Factors affecting the efficiency of Tullgren funnels

The questionnaires indicated that most soil ecologists use some form of Tullgren funnel and our studies show that the more sophisticated funnel-extraction methods are more efficient than mechanical methods and provide

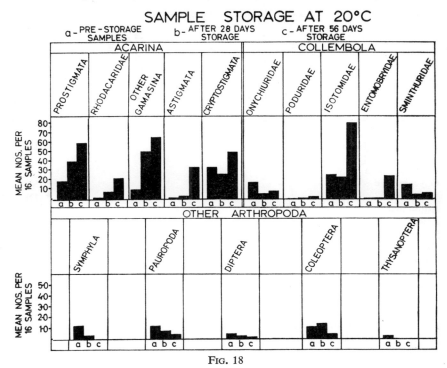

FIG. 18

better material for taxonomic study. It seemed important to study the use, mode of action and construction of funnels in detail, so as to recommend their best use and to suggest improvements in their construction. To do this, factors affecting the efficiency of the Rothamsted-pattern funnels were carefully compared usually with two sets of paired samples, extracting one set in the normal way and varying some factor for the other set. This was augmented by studies of the sequence of extraction of animals in four funnels by means of discs of plastic coated with 'Sticktite', slowly rotating (one revolution in 24 hours) under the outlet from the funnels.

(a) Orientation of the sample on the sieve

Some authors (Murphy, 1962) have suggested that animals can escape better when intact samples are inverted. Many animals live near the surface and these have a shorter distance to travel from an inverted than from an upright

TABLE V. Orientation of samples.

	Intact Inverted V. Crumbled Samples	Intact Inverted V. Intact Upright	Crumbled Samples V. Intact Horizontal	Double Sample V. Single Sample
Prostigmata	**	**	—	—
Mesostigmata { Gamasina	*	**	—	* (D)
Uropodina	0	—	—	—
Cryptostigmata	—	—	—	—
Onychiuridae	—	—	—	* (S)
Poduridae	—	*	—	* (D)
Isotomidae	—	*	—	** (D)
Entomobryidae	—	—	—	—
Sminthuridae	**	*	—	** (D)
Diptera	**	*	*	** (S)
Coleoptera	**	*	—	—
Thysanoptera	*	**	—	—

** First method significantly better 1 % level.
* First method significantly better 5 % level.
— No significant difference.
0 No animals.
(S) Single sample better.
(D) Double sample better.

sample. The following tests were therefore made: (i) sets of samples were extracted in upright and inverted positions, (ii) samples were crumbled apart and left intact and inverted during extraction, (iii) cores were extracted horizontally in the sieve and from crumbled samples.

The results are summarized in Table V. More animals were extracted from clay loam when the sample was placed on the sieve intact and inverted, than when crumbled or extracted upright.

(b) Size of sample

The depth of soil in the sieve may affect the efficiency of extraction, for animals may remain trapped in the upper part of the sample. If the sample is too shallow it may dry out too rapidly and animals may die before leaving the soil. A steep gradient of moisture and temperature through the sample seems to increase the efficiency of extraction but it is impossible to maintain a gradient with shallow samples.

(i) *Broken samples*

The effect of the depth of sample on efficiency of extraction was investigated by taking 16 sets of three soil samples 10 cm diam. × 5 cm deep from a small uniform area of pasture on clay loam soil. Two samples from each set of three were broken up and thoroughly mixed together then placed on the sieve of one funnel. The third was broken up and placed on the sieve in another funnel. This was repeated for all samples so that 16 funnels contained single samples and 16 had double samples.

The animals were extracted and sorted into categories then the numbers obtained from the double samples were halved before statistical analysis. Significant differences are summarized in Table V. Gamasid mites, podurid and isotomid Collembola and Psocoptera were extracted most efficiently from the double depth samples, but more onychiurid Collembola and Diptera larvae were recovered from single-depth samples.

Perhaps the more active animals move right through the greater depth of soil and respond better to a temperature and moisture gradient than less active ones, for the more sluggish fly larvae and deep-living Collembola were trapped in the double samples. Sections IV and V also demonstrate that dipterous larvae, and onychiurid Collembola are usually extracted better by flotation than by funnel methods. For most purposes the deeper soil sample appears to be preferable to a shallow one.

(ii) *Intact samples*

To compare the effect of depth of soil on the sieve using intact soil samples, 16 sets of three soil samples 10 cm diam. × 5 cm deep and 16 sets of three 5 cm diam. × 5 cm deep were taken as in the last section. One sample from each set was inverted on a sieve; another was sliced horizontally into two cores 2·5 cm deep and the two half-cores were inverted in two other funnels; the third was sliced horizontally into four sections each 1·25 cm deep and these placed in four more funnels. After the animals from a single original soil sample were extracted, they were bulked together and sorted into taxonomic categories. The sectioned samples tended to yield more Acarina but fewer Collembola than the unsectioned ones, but none of the differences was significant at the 5% probability level. More onychiurid Collembola and dipterous larvae were obtained from the sectioned samples than the unsectioned ones, confirming the results obtained with broken samples.

(c) Collecting fluids

Macfadyen (1962) postulated that the collecting fluid should be odourless, non-repellent and non-volatile and should kill the animals falling into it. The questionnaires indicated that most workers use ethyl alcohol although a few favoured water or other fixatives; picric acid was recommended for use in the Kempson extractor. To test the effect of water, ethyl alcohol and picric acid on the efficiency of extraction, sixteen sets of three 10 cm diam. $\times 5$ cm deep samples were taken from pasture on clay loam soil. Using Rothamsted type funnels one of each set of three was extracted into 70% ethyl alcohol* with 5% glycerol, one into picric acid and the other into distilled water. Table VI shows that many more animals were collected when extracted over picric acid and alcohol than over distilled water, except that dipterous larvae were recovered better over distilled water and picric acid than over ethyl alcohol.

The possibility that animals were escaping from the distilled water because they were not killed was tested by comparing two sets of sixteen samples, one collected into ordinary vials with distilled water and the other into vials with 'Fluon' painted on to the inside rim to prevent arthropods escaping, but there was no significant difference between the two.

It seemed that the alcohol and picric acid might attract animals and hence increase efficiency. This was tested by extracting four sets of sixteen samples, one into vials containing 70% alcohol and 5% glycerol, one into vials with distilled water but with small side-arms containing 70% ethyl alcohol, a third into distilled water and the fourth into distilled water containing a detergent, 'Teepol'. Significantly more Prostigmata, Gamasina, Onychiuridae, Sminthuridae, Thysanoptera and Hemiptera fell into tubes either with alcohol or into water with alcohol in the side arms than into water alone. Although more of other groups were also extracted in the tubes of alcohol, these increases were not significant. The only exception was dipterous larvae, numbers of which were significantly more in samples collected into distilled water. When ethyl alcohol was used as a collecting fluid, fewer animals were extracted with the Macfadyen canister apparatus (Macfadyen, 1962a) than when water was used. This was probably because in this apparatus the collecting fluid is completely enclosed just beneath the sample and the more concentrated vapours affect the efficiency of the extraction. For funnel methods ethyl alcohol or picric acid are probably best.

* Ethyl alcohol=Industrial methylated spirits, containing $\geqslant$ 5% wood naphtha.

TABLE VI.

	70% Alcohol + 5% Glycerol	Picric Acid	Distilled Water
Prostigmata	**	**	—
Gamasina	**	—	—
Cryptostigmata	**	**	—
Onychiuridae	—	**	—
Poduridae	—	—	—
Isotomidae	**	**	—
Entomobryidae	—	—	—
Sminthuridae	**	**	—
Diptera	—	**	**
Coleoptera	—	—	—
Thysanoptera	—	—	—

** = Means significantly greater than —. (P = 0·01).

VIII

Conclusions and recommendations of the best methods

The results of our experiments enable recommendations to be made about methods for extracting the different groups of arthropods from samples of soil. For productivity studies in the International Biological Programme, arthropod populations in selected sites will need to be accurately assessed and, because the productivity of ecosystems may be compared, methods for assessing populations should be standardized. Facilities, soils, fauna and research workers differ and rigid recommendations are impracticable. But it must be emphasized that before the arthropod populations in any particular site are properly assessed, extensive preliminary comparisons of the effectiveness of different extraction methods for that site, habitat, soil and climate should be made.

Our results indicate broadly the most preferable methods, but do not eliminate the need for tests in particular sites.

No single extraction method or size of sample is best for all groups of animals, and it may be necessary to use several methods and sample sizes. Alternatively where a single method is used for most of the estimations, its efficiency for each group of animals in that particular site should be investigated before it is used extensively. An efficiency correction factor can then be found for modifying estimates obtained with the technique.

TABLE VII. Recommended methods for extracting of soil arthropods

Soil type:	Peat			Clay or Loam			Sand		
Habitat: / Arthropod Group	Woodland	Pasture	Arable	Woodland	Pasture	Arable	Woodland	Pasture	Arable
Isopoda	AC	ABC	ABC	AC	ABC	ABC	X	X	X
Pauropoda	ABC	ABC	ABC	ABC	ABC	X	X	X	X
Symphyla	ABC	ABC	ABC	X	ABC	X	X	X	X
Diplopoda	ABC	ABC	ABC	ABC	ABC	X	X	X	X
Chilopoda	ABC	ABC	ABC	X	ABC	X	X	X	X
Prostigmata	AC	ABC	AC	AC	ABC	ACD	ABC	ABC	ABC
Gamasina	ABC	ABC	ABC	ABC	ABCDE	X	X	X	X
Uropodina	AC	ABC	ABC	AC	X	X	X	X	X
Astigmata	ABC	ABC	AC	ABC	ABC	AC	ABC	ABC	ABC
Cryptostigmata	ABC	ABC	AC	ABCF	X	ACD	ABCF	X	X
Pseudo-Scorpionida	ABC	ABC	ABC	ABC	ABC	ABC	X	X	X
Aranei	ABC	ABC	ABC	ABC	ABC	ABC	ABC	ABC	X
Onychiuridae	AC	ABC	AC	ACDEF	X	ACDEF	X	X	X
Poduridae	AC	ABC	ABC	AC	ABCD	X	ABCD	ABCD	ABCD
Isotomidae	ABC	ABC	ABC	ABC	ABC	ABCD	ABCD	ABCD	ABCD
Entomobryidae	ABC	ABC	ABC	ABC	ABC	ABCD	ABCD	ABCD	ABCD
Sminthuridae	ABC	ABC	ABC	ABC	ABC	ABCD	ABCD	ABCD	ABCD
Protura	ABC	ABC	ABC	ABC	X	X	X	X	X
Psocoptera	ABC	ABC	ABC	X	X	X	X	X	X
Thysanoptera	ABC	ABC	ABC	DE	DE	X	X	X	X
Hemiptera	ABC	ABC	ABC	X	X	X	X	X	X
Hymenoptera	ABC	ABC	ABC	X	X	X	X	X	X
Coleoptera	ABC	ABC	ABC	X	X	X	X	CDEF	CDEF
Diptera	ABC	ABC	ABC	X	X	X	X	CDEF	CDEF
Eggs and Pupae	DEF	DEF	DEF	DEF	DEF	DEF	DEF	DEF	DEF

A.   Simple funnel with heat.
B.   Kempson Infra-red extractor.
C.   Macfadyen air-conditioned funnel.
D.   Salt & Hollick flotation.
E.   Edwards & Heath flotation.
F.   Grease film extractor.
X.   All methods suitable.

Our study indicates that the best general method is a Tullgren funnel of the Macfadyen type, with a steep gradient of temperature and moisture through the sample. A simpler form of funnel of the Rothamsted pattern performed well for most groups of animals. To obtain eggs, pupae or animals in diapause, a mechanical or flotation method is essential. The way the funnels are used is also important. They should be kept in a well-ventilated constant temperature room and thoroughly cleaned between use. For most soils, cores should not be deeper than 5 cm or shallower than 2 cm, and they should be inverted in the sieve containers. The animals are best collected into either picric acid or 70% alcohol with 5% glycerol. Where possible, samples should be placed on the funnels the same day they are collected, but storage at 5°C for up to a week should not greatly affect the results.

Table VII lists the best methods for different groups of soil arthropods. In general, flotation methods are best with sandy soils but of less use in peat or muck soils with much organic matter. For most soils, funnels are most efficient for microarthropods but flotation is often preferable for the larger arthropods. Myriapods are extracted as well by funnels as by flotation from most soils, but their comparatively small numbers mean that large samples (10 cm diameter or greater) are needed. Gamasid and oribatid mites can be extracted efficiently by most methods, although funnels are usually preferable. Prostigmata and sometimes Astigmata are recovered very inefficiently by flotation, and funnels are recommended for them.

Most Collembola are more efficiently extracted by funnels than by flotation, except for the Onychiuridae which are better obtained by flotation. Other Insecta, particularly Diptera and Thysanoptera, are better recovered by flotation methods except from organic soils, but differences between flotation and the better funnel methods are seldom significant.

It must be emphasized that, to assess the numbers of soil arthropods accurately careful preliminary investigation is needed, and that adequate equipment and statistical procedure, very considerable labour and a thorough taxonomic facility are essential.

References

AUCAMP J.L. & RYKE P.A.J. (1964). A preliminary report on a grease film extraction method for soil microarthropods. *Pedobiologia*, **4**, 77–79.

BERLESE A. (1905). Apparecchio per racogliere presto ed in gran numero piccoli artropodi. *Redia*, **2**, 85–89.

BLOCK W. (1966). Some characteristics of the Macfadyen high gradient extractor for soil micro-arthropods. *Oikos*, **17**, 1–9.

DEBAUCHE H.R. (1962). The structural analysis of animal communities of the soil. *Progress in Soil Zoology*. Butterworth's, London. 10–25.

EDWARDS C.A. (1955). Soil sampling for symphylids and a note on populations. *Soil Zoology*. Butterworth's, London. 152–156.

EDWARDS C.A. (1967a). The effects of gamma irradiation on populations of soil invertebrates. *Proc. Second Nat. Symp. Radioecology, Ann. Arbor, Michigan*. (In press.)

EDWARDS C.A. (1967b). Relationships between weights, volumes and numbers of soil animals. *Progress in Soil Biology*. Verlag. Friedr. Vieweg & Schn., Braunschweig. 1–10.

EDWARDS C.A. & HEATH G.W. (1962). An improved method for extracting arthropods from soil. *Rep. Rothamsted exp. Stn*. 1962. 159.

EVANS G.O. (1951). Investigation on the fauna of forest humus layers. *Report on Forest Research for the year ending March, 1950*. Forestry Commission, London, H.M.S.O. 110–113.

FORD J. (1937). Fluctuations in natural populations of Collembola and Acarina. *J. Anim. Ecol*. **6**, 98–111.

HAARLOV N. (1947). A new modification of the Tullgren apparatus. *J. Anim. Ecol*. 115–121.

HARTENSTEIN R. (1961). On the distribution of forest soil microarthropods and their fit to 'contagious' distribution factors. *Ecology*, **42**, 190–194.

HEALY M.J.R. (1962). Some basic statistical techniques in soil zoology. *Progress in Soil Zoology*. Butterworth's, London. 3–9.

HUGHES R.D. (1962). The study of aggregated populations. *Progress in Soil Zoology*. Butterworth's, London. 51–55.

KEMPSON D., LLOYD M. & GHELARDI R. (1963). A new extractor for woodland litter. *Pedobiologia*, **3**. 1–21.

LADELL W.R.S. (1936). A new apparatus for separating insects and other arthropods from the soil. *Ann. appl. Biol*. **23**, 862–879.

MACFADYEN A. (1953). Notes on methods for the extraction of small soil arthropods. *J. Anim. Ecol*. **22**, 65–77.

MACFADYEN A. (1962a). Soil arthropod sampling. *Adv. Ecol. Res*. Academic Press, London and New York. 1–34.

MACFADYEN A. (1962b). Control of humidity in three funnel-type extractors for soil arthropods. *Progress in Soil Zoology*. Butterworth's, London. 158–168.

MORRIS H.M. (1922). On a method of separating insects and other arthropods from soil. *Bull. ent. Res*. **13**, 197–200.

MURPHY P.W. (1958). The quantitative study of soil meiofauna. 1. The effect of sample treatment on extraction efficiency with a modified funnel extractor. *Entomologia exp. appl*. **1**, 94–108.

MURPHY P.W. (1962). Extraction methods for soil animals. 1. Dynamic methods with particular reference to funnel processes. *Progress in Soil Zoology*. Butterworth's, London. 75–114.

NEWELL I.M. (1955). An auto-segregator for use in collecting soil-inhabiting arthropods. *Trans. Am. microsc. Soc*. **74**, 389–392.

SALT G. & HOLLICK F.S.J. (1944). Studies of wireworm populations. 1. A census of wireworms in pasture. *Ann. appl. Biol.* **31,** 52–64.

SATCHELL J.E. & NELSON J.M. (1962). A comparison of the Tullgren-funnel and flotation methods of extracting acarina from woodland soil. *Progress in Soil Zoology.* Butterworths, London. 212–216.

SKELLAM J.G. (1962). Estimation of animal populations by extraction processes considered from the mathematical standpoint. *Progress in Soil Zoology.* Butterworth's, London. 26–36.

TUKEY J.W. (1953). *The Problem of Multiple Comparisons.* Princeton University, Princeton, N.J.

TULLGREN A. (1918). Ein sehr einfacher Ausleseapparat fur terricole Tierformen. *Z. angew, Ent.* **4,** 149–150.

ULRICH A.T. (1933). Die Makrofauna der Waldstreu. Quantitativ Untersuchungen in Bestanden mit guter und schlechter Zersetzung des Bestandesabfalles. *Mitt. Forstw. Forstwiss.* **4,** 283–323.

VALLE A. (1951). Su un nuovo sistema di selezione particolarmente adatto per la raccolta in massa dei microarthropodi. *Boll. Zool.* **18,** 299–304.

WOOD T.G. (1965). Comparison of a funnel and a flotation method for extracting Acari and Collembola from moorland soils. *Pedobiologia,* **5,** 131–139.

11

Mites

P. BERTHET

(I) Introduction

Up to the present, the study of energy flow through populations or communities of mites has been the object of only a limited number of investigations. It seems that acarologists have been trying for a long time to solve the problems of sampling, extraction and population census. Moreover, the great number of edaphic species and difficulties of identification have definitely impeded the progress of knowledge in this sphere. Further the great specific diversity in a single site has led ecologists to compare on the one hand the ecology and ethology of different species and, on the other, the populations of different biotopes.

This work is an attempt to summarize the numerous methods which have been used for studying the productivity of edaphic organisms. Thus we hope to promote new research which will enable us to delineate and quantify the activity of mites in their environment and, finally, to understand better the position of these organisms within the structure of communities.

(II) Estimation of density and biomass

Any population or community energetics study implies from the start an estimate of the density of the organisms concerned. For the edaphic mites, such an estimate is made by counting individuals in a certain number of sample units. Therefore, as a preliminary, one must decide the dimension and number of sample units to be taken, the frequency of sampling and the method of extraction to be used.

After counting the individuals, the density D is generally expressed per unit area (square metre) and is estimated by

$$\hat{D} = \frac{\bar{x}}{S}$$

where $\bar{x}$ is the mean number of individuals per sample unit and S the area (in m²) of a sample unit.

(1) The sample

(a) Size

Samples are most often taken at a constant volume and area by means of a cylindrical corer. Various kinds of corer have been described (Coile, 1936; Alexander & Jackson, 1955; Cohen, 1955; Macfadyen, 1953, 1961; O'Connor, 1957; Murphy, 1962; Von Torne, 1962; Vannier and Vidal, 1965; Vannier and Cancela da Fonseca, 1966). They generally have an area of between 10 and 20 cm² and penetrate the soil to a depth of 2–6 cm. Such corers are suitable for relatively compact soil, but have to be used carefully in sampling forest litter. In this case, sample units should be obtained either by keeping the volume constant (Van der Drift, 1950; Lebrun, 1965), or by taking samples of a constant area using a quadrat or frame (Gérard, 1967).

In general, the area of the sample unit is known with a fair degree of accuracy, and the precision of the estimate of density is determined essentially by the precision of the estimate of the mean.

It is known that for a random sampling the precision of the estimate of the mean depends on the square of the number of observations. The number of units to be taken in order to attain a certain accuracy in estimating the mean could be calculated, as Healy (1962) suggests, by the equation:

$$m = \bar{x} \pm t_1 - \alpha/2; n - 1 . \frac{s}{\sqrt{N}}$$

where t is the value of Student's table for a level α of selected probability and s the estimated standard deviation of the population. Under these conditions, the number of units to be taken is approximately

$$N = \left(\frac{ts}{p\bar{x}} \right)^2$$

where p is the required accuracy expressed as $p = \dfrac{m - \bar{x}}{\bar{x}}$ (Southwood, 1966).

Such reasoning implies, however, that the standard deviation has already been estimated with sufficient precision in previous experiments. If one only has a rather rough estimate of the variance, it is preferable to decide how many units to take by referring to other methods, like those of Harris *et al.* (1948) or of Stein (1945).

The use of these methods presupposes that the variable studied is normally distributed, which is never the case for samples of animal populations. Under these conditions, in order to estimate how many units to take, Cancela da Fonseca (1966) recommends effecting a transformation of the variable in order to normalize the distribution. This can lead to the determination of the number of units to be taken to obtain a given accuracy in estimating the mean of the transformed variable, and can prove useful in problems of comparison between populations. Such a method is only indirectly useful in problems of estimation which arise in the study of productivity.

On the other hand, if the number of individuals per unit area is a Poisson variable, the estimate of the mean depends solely on the total number of counted individuals x_0, of which the confidence interval with a coefficient α is given by the equations (Hald, 1952):

$x_{max} = 1/2 \; \chi^2 1 - \alpha/2$ with $2(x_0 + 1)$ degrees of freedom
$x_{min} = 1/2 \; \chi^2_{\alpha/2}$ with $2x_0$ degrees of freedom.

For example, if within 20 sample units 10 individuals of the same species have been observed, the max. and min. values up to 0·05 of this total are:

max. value $= 1/2 \; \chi^2 0·975 ; 22 = \frac{1}{2} . 36·78 = 18·39$
min. value $= 1/2 \; \chi^2 0·025 ; 20 = \frac{1}{2} . 9·59 = 4·795$

Thus the mean number of individuals per sample unit is 0·5 and is comprised (at the 5% level) between 0·92 and 0·24. The size of the interval of confidence is therefore 0·92—0·24 = 0·66, that is 1.34 times the mean.

On the other hand, if there is a total of 50 individuals in the 20 sample units, one obtains for the same threshold the max. and min. values of 62·15 and 37·11, respectively; the size of the interval is no more than 0·5 times the mean. Therefore, it is possible to estimate in this way the number of individuals to be counted in order to obtain the required precision.

The use of this method presupposes that the population studied is distributed in a homogeneous way within its habitat (Poisson distribution). In fact, many studies (e.g. Hartenstein, 1961; Debauche, 1962; Nef, 1962; Berthet and Gérard, 1965; Rapoport, 1966; Gérard, 1967) have shown that edaphic mites generally show an aggregated repartition in their environment: the distribution of individuals in sample units is generally aptly described by the negative binomial distribution. It is known that for such a distribution the variance σ^2 is

$$\sigma^2 = m + \frac{m^2}{k}$$

where m is the mean number of individuals per sample and k is a parameter which can be considered as a measure of the heterogeneity of spatial distribution (Berthet and Gérard, 1965) and seems, under normal sample conditions, to be practically independent of m (Gérard, 1967).

If the density D of the population is estimated at the start from a preliminary sample of N independent units of area S, and if $\bar{x}$ is the mean number of individuals per unit, the estimator of D is

$$\hat{D} = \frac{\bar{x}}{S}$$

Thus, the relative error of this estimator is

$$E_d = \frac{\sigma_d}{D} = \sqrt{\frac{1}{NS}\left(\frac{1}{D} + \frac{S}{k}\right)}$$

(Berthet and Gérard, 1970).

It appears, therefore, that, if the total area of the sample remains constant (NS=constant), the relative error will decrease as the density increases. For a very abundant species, $1/D$ can become negligible by comparison with S/k; under these conditions E_d will be more or less equal to $1/\sqrt{Nk}$ and the precision of the density estimate will depend directly on the number of units taken.

On the other hand, the relative error increases with the degree of aggregation $(1/k)$ of the population. In contrast, for a homogeneous distribution of individuals corresponding to a Poisson type of sample distribution, the value $1/k$ is zero and the relative error is $E_d = (NSD)^{-\frac{1}{2}}$. For non-aggregated populations, the precision of the estimated density will, therefore, depend solely on the number of individuals counted, as was quoted above.

Thus, if the relative error depends on the characteristics of the population, it also depends on the methods of sampling. For a given population (where D and k are constants) and a total sample area, it is possible to reduce the relative error by reducing the surface area of the sample unit.

For a total area examined of $0\cdot1m^2$, Figure 1 shows relative error curves of 10, 20, 30, 40 and 50% in terms of the radius of the corer, and the population density, with three values for k. Thus, for a population of 2,000 individuals per sq. metre, sampling with a corer of 6 cm radius will give a relative error of 40 to 50% if $k=0\cdot5$, 30 to 40% if $k=1$, and 20 to 30% if $k=2$; on the other hand, if the corer has a radius of 3 cm, the relative error will be reduced and

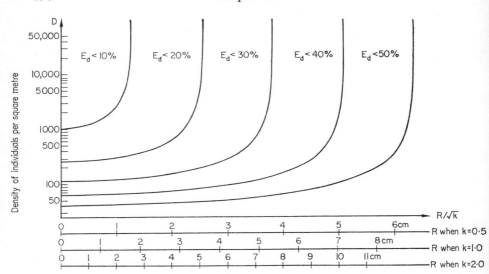

Figure 1. Curves of relative error in the calculation of the density in function of D and $R/\sqrt{k}$ for a total surface area examined of 0·1 m².
On the base line the scale of relative error for different values of k.

will be in the order of 20 to 30% if $k = 0·5$ and less than 20% if k is equal to or more than unity. If the total area is not 0·1m² but NSm², the relative error values shown on the graph must be multiplied by

$$\sqrt{\frac{0·1}{NS}}$$

It appears, therefore, that, in order to estimate density with the maximum possible precision, it is always an advantage (whilst one is counting) to take a large number of sample units of small area.

It is up to the experimenter to decide which is the lowest area admissible for a sample unit. In fact, the use of a corer of small diameter can complicate the determination of the true sample area taken. In addition, the cutting effect and the pressure exerted by small corers can damage the organisms and thus appreciably reduce the estimate of the density. Thus, the efficiency of the extractor used in the classic way for small samples can become impaired.

(b) *Frequency of sampling*

In order to study the ecology of edaphic mites and in particular to 'correlate' the modifications of environment to those of the animal, or to compare the fauna in different environments, it is usual to carry out sampling in correspondence with the different phenological phases. However, for productivity studies, it is better to make a continuous measurement of density throughout the year. In this case, frequent sampling, even if each sample is not composed of more than 2 or 3 units, is preferable to a larger, but infrequent, sample. Thus, in a previous study (Berthet, 1964), the census of a population of edaphic Oribatei in an oak plantation was based on monthly samples of 25 units each, a total of 300 units for the year (Lebrun, 1965). For the calculation of metabolic activity, we were led to suppose that the densities observed at a given moment reflected the state of the populations during the 15 days between each sampling date. In fact, for this work, it would have been preferable to use, for example, weekly observations based on 5 or 6 sample units. Here again, a compromise must be found between the demands of theory and the practical possibilities of sampling in the field, distances to be covered, and the man-power available.

(2) Methods of extraction

A review of all the methods put forward for extracting edaphic mites from their substratum would be too wide a subject for this chapter. The reader can refer to the excellent summary published by Macfadyen (1962) on this subject, and to the comparative study of the efficiency of different types of extractor, published in another connection in this work (Edwards and Fletcher, Chapter 10).

(3) Determination of biomass

Biomass means the weight of living material of one species or group of species per unit area, or sometimes per unit volume. It is determined by summing, for each species, the products of the densities and mean weight of an individual.

Relatively few authors have tried to determine the weight of different species of mites. A few determinations have been made by weighing groups of individuals on analytical scales (Macfadyen, 1952; Zinkler, 1966). The best method, however, consists of making individual weighings using quartz wire torsion balances (Engelmann, 1961), or electrobalances (Berthet, 1963, 1964).

The weight of adult animals belonging to about 30 species of Oribatei was determined in this way, and the weight of individuals of about 40 species of less common Oribatei was estimated by comparison with individuals whose weight was known (Berthet, 1964).

For further work (Berthet, 1967), a multiple regression equation linking the logarithm of weight with the logarithm of length and the logarithm of width, was calculated for about 30 species of Oribatei which had been weighed previously (the *Phthiracaroïdea*, being very different from other species of Oribatei, were not considered). The regression equation obtained was:—

$$P = L^{1.58} \cdot l^{1.45} \cdot 10^{-6.61}$$

where P is the weight in μg, L and *l*, the length and width in microns.

However, it must be observed that for the calculation of the regression equation, the lengths and widths are not those of the individuals weighed, but come from data in the literature.

Suspecting size variations in individuals of the same species according to the biotope from which they originate, Lebrun (1968) determined the length, width and weight of numerous individuals belonging to about 20 species, and under these conditions, he obtained the equation:

$$P = L^{1.53} \cdot l^{1.53} \cdot 10^{-6.67}$$

In addition, this author showed that it is possible to obtain a better estimate of the weight of Oribatei by classifying different species according to morphological types. Thus, for the 'Achipteriformes' the equation is:

$$P = L^{2.09} \cdot l^{0.93} \cdot 10^{-6.67};$$

for the 'Carabodiformes'

$$P = L^{1.62} \cdot l^{1.40} \cdot 10^{-6.56}$$

and for the 'Nothroïformes'

$$P = L^{2.09} \cdot l^{0.84} \cdot 10^{-6.44}.$$

Here we are only dealing with more or less accurate equations of prediction; if there is a microbalance available, it would be preferable to make direct weighings.

(III) Determination of energy flow

Ideally, the establishment of an energy budget requires, in addition to estimates of density and biomass, estimates of the rates of ingestion, assimilation and egestion, of respiratory activity, birth and death rates. In fact, the fundamental equation of secondary productivity is $C = A + F$, indicating that the quantity of energy ingested (C) is equal to the sum of the quantity of energy assimilated (A) and of energy egested (F).

The energy assimilated is partly dissipated by the respiratory process (R), the rest of it being used in the processes of production of new materials (P), (A = R + P). The quantity of energy produced per unit time depends on the variations in biomass during the period under consideration (generally a year) and on the turnover rate of the population; the energy produced will thus depend on the birth and death rate.

Up to the present time, very few ecologists have tried to determine, for mites, the values of the different terms in the equation. Only Engelmann (1961) attempted to estimate each term independently; on the other hand, a few authors (Bornebusch, 1930; Berthet, 1963, 1964, 1966) have measured the respiratory activity of mite populations, and, from this, have tried to estimate the values of the other terms of the equation.

(1) Measurement of the food consumption: determination of the quantity of food ingested, assimilated and excreted.

(a) *Methods*

The quantitative determination of the feeding activity of edaphic mites has not been studied very much up to the present. Two approaches have been used: gravimetry and radioisotope methods.

Working by direct weighings, Engelmann (1961) tried to determine the quantity of food ingested by Oribatei in the laboratory. Berthet (1962) preferred to proceed by planimetry; discs of several millimetres diameter of different plant materials were offered to a large number of *Steganacarus magnus* (Oribatei). After a certain time, what remained of the food was photographed; on a photographic enlargement, the area of the leaf eaten can be measured by planimetry. Knowing area/dry weight relationship of each of the discs, one could estimate the weight of food eaten per individual per day. Unfortunately, a purely superficial attack on the dead leaves cannot be detected by planimetric measurement. Similarly, Murphy (1956) and Berthet (1964) have been able to determine approximately the quantity of food egested by the animals in the laboratory: in isolated laboratory populations of *Steganacarus magnus* it was possible to count the faecal pellets produced per unit time, to estimate the mean dimensions, and to deduce the mean amount eliminated per day.

The quantitative study of the feeding habits of Oribatei by the use of radio-isotope tracers was started by Engelmann (1961) and Berthet (1964). The first of these authors presented the animals some food containing a radioisotope (e.g., glycin ^{14}C incorporated in a yeast) with a known radioactivity rate. After a certain time the animals were killed and their radioactivity determined; the result was then related to the quantity of yeast in the animal's body. From this information the author deduced the daily rate of ingestion. Similarly, the determination of the radioactivity rate of animals which could have digested some food before being killed, enabled the author to estimate an assimilation rate. Engelmann (*op. cit.*) was very discreet about the technical details and the methods of calculation which he used; similarly, he did not specify either the species of Oribatei or the number of measurements carried out.

In order to label dead leaves with a radioactive substance, Berthet (1964) induced the precipitation of $Ba^{35}SO_4$ into the very heart of the leaf tissue, and after determining its radioactivity rate, presented this food to *Steganacarus magnus* in the laboratory. Under these conditions, it was observed that the mean radioactivity rate of individuals increased during the first 24 hours and remained at a constant level during the following days. From the outset of these observations, one could thus estimate the amount of food present in the digestive tract: this corresponded to the minimal amount which the animals are liable to eat in 24 hours.

(b) *Discussion*

There are several remarks to make about these methods

(1) The observations made on the feeding habits of laboratory animals must be accepted with great caution, as they can only give a very rough picture of the food requirements of animals in the field; the general laboratory conditions are always different from natural conditions and there is a risk that the animals' behaviour will be considerably modified. The lab. conditions themselves must be carefully controlled. Rodriguez *et al.* (1967) observed that protonymphs of *Tetranychus urticae* (Acarina) had an ingestion rate 33% higher when the individuals were clinging to a support, ventral side upwards, than when they were lying on the bottom of a petri dish, dorsal side upwards! Similarly, these authors showed that the adults of this species have a higher food consumption in green or yellow light than in red or white light.

(2) Food consumption is liable to vary considerably with the quality and quantity of food given to the animals. Berthet (1964) showed that

TABLE 1. Amount of food eaten (dry wt.) by *Stegana-carus magnus* (average of 70 individuals)— different foods.

Corylus avellana	8·7 µg/ind./day
Carpinus betulus	8·3 µg/ind./day
Quercus robur	8·5 µg/ind./day
Fagus silvatica	8·2 µg/ind./day

Steganacarus magnus, fed solely on hazel leaves, ingested 1·67 times more food than when it was fed mainly on beech leaves (Table 1).

Similarly, in order to study the feeding behaviour of *Prodenia eridania*, Soo Hoo and Frankel (1966) fed these insects with 18 different species of plants. They found that the 'coefficient of digestibility,

$$(\frac{\text{dry wt. ingested} - \text{dry wt. excreted}}{\text{dry wt. ingested}}),$$

which corresponds to what Wiegert (1964) calls assimilation/consumption, varied, according to the food, between 36% and 76%. The efficiency of the conversion of digested matter (dry wt. gained by the animal/dry wt. of digested food), i.e., the production/assimilation yield, can also vary from 16 to 50%. These same authors showed that for a type of food favourable to the growth and development of *Prodenia* larvae, there seems to be a negative correlation between the amount of food consumed and the efficiency of its use. Similarly, Hubbel *et al.* (1965) using radioisotope methods noticed that Isopoda in the lab. had an appreciably higher assimilation/consumption yield when food was given to them in small amounts than when they were continually given an excessive amount; in captivity, they had an assimilation rate of between 11 and 13%, whereas in the wild the rate was 53 to 75%; on the other hand, the ingestion rate was two or three times greater in the lab. than it was in natural surroundings.

(3) One must also mention the very ingenious method used by Phillipson (1960) in his study of the consumption of *Mitopus morio*. This author made use of the fact that the above mentioned Phalangida, when feeding on a Heteroptera (*Calocoris*), produced green faeces. Knowing the weight of the prey, and being able to identify the excrement produced from this food, he could estimate the assimilation/consumption ratio. With this information, and with the measurement of the amount of faeces produced by the animals after capture, he was able to estimate the amount of food consumed. How-

ever, it must be noted that such a calculation was based on the hypothesis that the assimilation efficiency remains constant, whatever the prey.

Phillipson's method has been used for the study of the feeding habits of herbivorous organisms and is now being used by Bocock and Heath (1967) in the study of the feeding habits of *Glomeris marginata*. It seems that such an approach could be tried in connection with the feedin habitsg of mites.

(4) The use of radioisotope methods in the study of feeding habits, only recently developed, seems to be full of possibilities, as these techniques very often permit an approach which is non-destructive of animal populations and, in particular, permits the examination of the behaviour of organisms in natural conditions. On the qualitative level, thanks to these methods, it is possible to tackle the study of food chains by marking specific plants and analysing the degree of dependence of the animal on the marked plant. Various isotopes have been used to this end: ^{32}P (Pendleton and Grundman, 1954; Odum and Kuenzler, 1963; Paris and Sikora, 1965, 1967; Marples, 1966; Williams and Reichle, 1968), ^{131}P (Grösswald, 1963), ^{83}Sr and ^{57}Co (McMahon, 1963), ^{131}Ca for the different studies carried out on the experimental site at Oak Ridge, Tennessee (Reichle and Crossley, 1965, 1967a, 1967b) and ^{86}Rb (Williams and Reichle, 1968).

In addition to this qualitative information, radioisotope methods have in recent years supplied quantitative information on the feeding activity of different soil arthropods (*c.f.* works of Hubbel *et al.*, 1965; Crossley, 1966; Paris and Sikora, 1967). These quantitative studies are based on a theoretical model suggested by Davis and Foster (1958) and completed and discussed in a remarkable way by Paris and Sikora (1967) and by Reichle and Crossley (1967a).

One can say, very roughly, that as soon as animals fed with a radioactive substance reach a constant rate of radioactivity, the flow of the isotope through the population can be described by the equation $ra = Q_e k$, where r is the ingestion rate of the radioisotope, a the fraction assimilated, k the rate of elimination, and Q_e the concentration of the isotope in the animals' bodies when equilibrium is reached. The rate of elimination k is determined by measuring the time taken to reach equilibrium and, therefore, depends on the biological half-life of the isotope used. As Paris and Sikora (1967) point out, it is possible that various physiological processes converge at different rates in the elimination of the isotope; in this case the elimination curve is composite and must be analysed as such.

Note that the study of feeding activity by the radioisotope method assumes that the absorption of the radioisotope is not selective and that the animals are unable to distinguish marked food from unmarked food. In the study of feeding habits of Oribatei, this is proving a considerable difficulty, as dead leaves made radioactive by chemical means do not appeal to the animals in the same way (Berthet, 1964). Thus it is necessary to resort to methods of biological marking, like that used at Oak Ridge, where injections of radio-cesium-137 were carried out on trees in one patch, or like that used by Gifford (1967), who is growing Scotch pines in a greenhouse in the presence of radio-active CO_2. Such methods suggest that considerable costs are involved, and it will be several years before there are any results. It is feared, unfortunately, that these difficulties will limit the use of radioisotope methods in the study of the feeding habits of mites to a few specialized centres.

(2) Measurement of oxygen consumption

Since the work of Bornebusch (1930), the study of respiratory activity of soil organisms has developed quite appreciably. On the one hand, the comparison between metabolic activity of the different populations, rather than the comparison between their density or biomass, often forms a significant basis for studying the impact of the animal on its environment. This approach was used, for example, by Berthet (1967), where the author estimated the amount of oxygen consumed by the different species of Oribatei present in 13 forest biotopes. He determined, for each biotope, the total amount of oxygen consumed per year by the Oribatid community and the relative importance of each of the species in the metabolism of the community. For such a comparison, it was assumed that the ratio assimilation/respiration or consumption/respiration was constant from one species to another.

On the other hand, as the production of mites is fairly low (slow growth, eggs generally small and rather scarce), the energy dissipated during respiratory processes represents the essential part of the energy assimilated (Healey, 1967). Engelmann (1961) reached the conclusion that for soil arthropods, only 4% of the energy assimilated is used for production. In these circumstances, the measurement of respiratory activity of mite populations can provide a satisfactory estimate of the amount of food assimilated, and, knowing the assimilation/ingestion ratio, it is possible to calculate the amount of food consumed.

(a) *Laboratory methods*

The few authors who have tackled the problem have tried to measure, in the laboratory, the oxygen consumption of a few individuals belonging to different species. The small size of the individuals and their weak metabolic activity limit the number of methods of measurement in this sphere.

The Warburg apparatus was used in a crude fashion by Bornebusch (1930). This author was only interested in a few species with large individuals; in addition, he put a great number of individuals into each flask. More recently, this method was used by Zinkler (1966), who had some Warburg microflasks made by the firm B. Braun (Melsungen, Germany), with an approximate volume of 0·8ml, which enabled him to detect a minimal consumption of 0·05 μl of oxygen.

The present method which seems best adapted to the study of the metabolism of mites, especially when considering small individuals, is that of the 'Cartesian diver respirometer' of Linderström-Lang (1943), Holter (1943) and Zeuthen (1950). It was used by Berthet (1964) exactly as described by these authors. This technique has been improved in different ways: Løvlie and Zeuthen (1962) suggested placing the 'Cartesian diver respirometer' in a density gradient; thus manometric measurement is replaced by measurement of the distance covered by the diver, and by automatic photographic recording, it is possible to follow the consumption of oxygen. On the other hand, Klekowski (personal communication) used KOH as a floating liquid, which simplifies considerably the filling of the 'Cartesian diver respirometer.'

Finally, one must mention the electrolytic microrespirometer of Macfadyen (Macfadyen, 1961; Fourche, 1963, 1967). This apparatus has enabled Fourche to follow variations in oxygen consumption during the pupation of *Drosophila melanogaster* when consumption falls to 1·5 μl O_2/pupa/hour. In order to use this microrespirometer for the study of the metabolic activity of edaphic mites, it seems that the sensitivity of the apparatus would have to be increased approximately ten-fold.

(b) *Discussion*

1. It is evident that the value of oxygen consumption measurements of animals kept in a Warburg flask or in a 'Cartesian diver respirometer' only gives a very rough approximation of the amount of oxygen consumed under natural conditions. However, we agree with Nielsen's comment (1949): 'it has often been claimed that most determinations of O_2-consumption were performed under unnatural conditions, either the environment or physiological state of the object deviating from those found in nature. Of course, this point of view may be correct; in most cases, however, one has to choose between

measurements under circumstances deviating more or less from the natural ones or no measurements at all.'

2. The influence of temperature on the respiratory activity of mites seems considerable; Berthet (1964) studied the effect of this factor on individuals belonging to 16 species of edaphic Oribatei whose oxygen consumption was measured at 0·5, 10, 15 and 25°C. Between 5 and 15°C, the calculated value of Q_{10} was of the order of 4, and hence considerably higher than that forecast by Krogh's curve.

It is certainly regrettable that the influence of temperature rhythms upon the metabolism of mites has not been studied. In fact, it is feared that by keeping the animals in a constant temperature, one is partially inhibiting their metabolic activity, as shown by the results of Gromadska's experiments on insects (Gromadska, 1949) or on different Annelidas (Gromadska and Przybylska, 1960; Gromadska, 1966).

In order to tackle this problem, a respirometer would have to be used which would allow observations over long periods, with changing temperatures; at present, it seems that Macfadyen's microrespirometer would meet these demands.

3. The physiological state of the animals used for these measurements is very important. The use of laboratory animals does not seem to be indicated, as the lab. conditions could appreciably modify their behaviour. On the other hand, by placing freshly collected animals in a respirometer, one produces a thermal shock which can be quite considerable; the animals undergo a thermal shock when they pass through the extractor, and then, quite often, a second one when they are placed in the respirometer. This can cause appreciable modifications in metabolic activity (*cf.* for example, Agrell, 1947). Here again, measurements should be made of oxygen consumption over long periods.

In addition, the probable effect of a seasonal factor should not be ignored; independently of the temperature, the metabolism and the value of Q_{10} observed in poïkilotherms can fluctuate quite considerably according to the time of year under consideration, as shown, among others, by Bert (1951), Vernberg (1952), Eliassen (1953), Krog (1954), Rao and Bullock (1954), Florkin (1960) and more recently by Phillipson (1967).

Finally, Phillipson's (1963) studies show the necessity of taking into account sex, degree of development of gonads, or age of individuals. On this subject, it is very regrettable that, at the present time, we do not have very extensive information on the oxygen needs of the larval and nymphal forms of mites, for in their natural habitat, the immature forms constitute a large part of mite communities and seem to have a considerably higher metabolic activity than the adults.

To conclude this paragraph, may we quote Phillipson (1966): 'If respiratory data are to be used in conjunction with population data to calculate population metabolism, then it is essential that measurements of metabolic activity be made on all life stages of a single species, throughout the course of a year. Further, because temperature influences metabolic rate, it is desirable to measure respiratory rate at the same temperature as the natural environment. . . . Finally, whenever possible, measurements of respiratory rate should be taken over a period exceeding 24 hours, thereby counteracting differential respiratory rates due to either diurnal or nocturnal activity. . . .

Studies of the respiratory metabolism of populations are in their early stages, but no doubt the coming decade will see great advances made.'

(c) *Estimate of annual respiratory metabolism*

The integration of laboratory measurements with census and biomass data presents incontestable difficulties and this problem has been the subject of various discussions (*cf.* Phillipson, 1963; Wiegert, 1967; Macfadyen, 1967).

One of the first methods, frequently used since Bornebusch (1930) and Nielsen (1949), consists in multiplying the mean rate of oxygen consumption (i.e., O_2-consumed per hour/weight of an individual) obtained for a species or group of species, by the mean biomass of the group under consideration. This corresponds to Phillipson's first method (1963) ($mm^3 \, O_2/mg/h \times mg/m^2$)

As not all the individuals of the population have the same weight, this method seems very approximate. In fact, does not the calculation of the oxygen consumption rate, that is, the relation between the amount of oxygen consumed and the weight, imply the existence of a linear relationship between these two variables? Many authors, mainly Zeuthen (1947, 1953) and more recently (among the ecologists interested in the study of energy flow), Engelmann (1961), Berthet (1964), Wiegert (1965), Zinkler (1966), etc., have observed a relationship of the type O_2-consumed $= kW^b$, (where k and b are constants) between the metabolic activity and the weight (W). When interspecific comparisons are made, b is generally in the region of 0·7 or 0·8. Under these conditions, it is evident that by multiplying the consumption rate by the biomass, one is systematically slanting the estimate of the amount of oxygen consumed by the animal.

Another method which takes into account the weight of the individuals consists in determining empirically the relationship which links the oxygen consumed with weight and in tracing the line which approximately joins these variables. In this way, one avoids forming the hypothesis of a linear relationship between the oxygen consumed and the logarithm of the weight. By diagrammatic interpretation, it would thus be possible to predict the oxygen needs of an individual of known weight. This method would be particularly appropriate when the individuals whose oxygen consumption has been measured, show very different degrees of development (for example, see Fig. 2 of Phillipson's work (1963)).

In other cases, it would be useful to subdivide the population in terms of the degree of development of the gonads and to calculate metabolic activity on this basis (Phillipson, 1963).

Similarly, the effect of temperature should be included in the calculation of the metabolic balance. The ideal method would be to select, during each sample, a few individuals whose oxygen needs would be determined at the temperature prevailing at that moment. On this basis, the metabolism of the total population could be calculated for each period and, eventually, for the whole year. Such a procedure cannot generally be continuous and authors would have to restrict themselves to introducing a corrective factor in order to take into consideration the effect of temperature.

Bornebusch (1930), for example, based the calculation of annual metabolism on the annual means of density and temperature. Nielsen (1949) considers that the Nematodes do not exert an appreciable activity except during a hundred or so days per year; he estimated that the temperature at this period was 16°C. As the respirometric measurements were carried out at this temperature, the experimental results were used without correction. Later, this same author (Nielsen, 1961) again analysed these data by taking into account a mean monthly temperature curve and by correcting, with Krogh's curve, the data for oxygen consumption observed at 16°C in the laboratory.

Engelmann (1961) also used Krogh's relationship of respiration/temperature and, basing his calculations on the mean monthly air temperature of the experimental region, he estimated that the annual metabolism corresponded to an activity over a period of 120 days at 25°C, at which temperature the measurements were taken.

On the other hand, Wiegert (1967) corrected the oxygen consumption calculated for each period by means of a curve of oxygen consumed/temperature which he determined experimentally in the laboratory.

Finally, Berthet (1964, 1967) followed the variations of temperature in the different environments studied by means of an integrating system (Berthet, 1960) and corrected the lab. data to natural conditions by basing his calculations on an oxygen-consumed/temperature relationship obtained experimentally when studying the respiratory behaviour of 16 species of adult Oribatei. In order to determine the metabolic balance of Oribatei communities, which usually consist of more than 50 species, he established a general relationship linking the oxygen consumption to the weight of the individuals and the temperature to which they are subject:

$$y = 18 \cdot 054 + 0 \cdot 70 \ W - 0 \cdot 487 \ Z$$

where $y = \log_{10}$ of the oxygen consumption per individual per day, in 10^{-3} µl.
$W = \log_{10}$ of the individuals in µg.
$Z = 10^4/T$, where T is the absolute temperature.

Thus, for the different species and different periods considered, it is possible to estimate the respiratory balance of each community.

(b) Measurement of production

In order to make use of all the parameters of the energy flow equation, it is necessary to be able to estimate the amount of energy or material produced by the populations being studied. This implies knowledge of the birth and death rates for individuals belonging to different age groups. It is evident that information obtained on these subjects in the laboratory are definitely not applicable to natural conditions. With reference to mites, only Engelmann (1961) has put forward certain estimates. After a few observations concerning the lab. conditions, he put forward the idea that the Oribatei lay on average 1 egg every 10 days, that laying only occurs for 6 months, and that the population renews its biomass annually; in addition, the author makes estimates of the calorific value of the eggs and the adults. Under these conditions, the measurements of the annual production can only be approximate.

In a more recent study, Engelmann (1966) established a prediction relationship linking the net productivity and the respiratory metabolism. For poïkilotherms, this equation is

$$\log P = \frac{\sum\limits_{1}^{n} \log R - 0 \cdot 62}{0 \cdot 86}$$

where P is the net production in Kcal/m²/year, R is the respiratory metabolism in the same units and n is the number of poïkilotherm species in the community. This regression equation is an attempt to generalize the results of different studies, and has been based on the values of energetic balances obtained in different communities by Odum and Smalley (1959), Smalley (1960), Engelmann (1961), Odum *et al.* (1962) and Wiegert (1964).

To what extent is it permissible to group the results of observations carried out on such different organisms as *Littorina*, Nematoda, Orthoptera or Acari? Of course, this is a debatable point, but the regrouping of observations by regression would lead one to belive that the production/respiration relation is fairly constant from one poïkilotherm to another. Even so, there is still much work to be done to verify this hypothesis.

(IV) Conclusions

Having considered the different methods used by ecologists, it must be remembered that acarologists are still a long way from having a very precise idea of the energy flow which runs through populations and communities of mites.

In order to fit these organisms into a trophic scheme there is still a lot of information needed on the biology of these animals. In particular, the feeding regime of many edaphic mites is still relatively unknown: if some are predators, others are saprophagous, phytophagous, mycophagous, algophagous, etc. . . . The reproductive cycles in natural conditions, and mortality, have not yet been sufficiently studied. It must be hoped that, during population censuses, ecologists will try to distinguish the different phases of the organisms and that, with the aid of statisticians, they will try to construct mathematical models from which they will be able to make estimates of natality and mortality in nature.

Finally, there should be close collaboration between all pedobiologists in their study of the impact of mites on their biotope within the biocoenosis.

Acknowledgements

This work was carried out in the Laboratory of Animal Ecology directed by Prof. H. R. Debauche to whom I am thankful for his advice and encouragement.

I also wish to express my gratitude to Drs. G. Gérard and P. Lebrun in respect of the numerous hours spent by them in discussing and improving the contents of this chapter; perhaps they have left to me only those errors and omissions which have crept in.

Finally my thanks to Miss Susan Lee who translated the chapter from the original French.

References

AGRELL I. (1947). Some experiments concerning thermal adjustment and metabolism in insects. *Arkiv. Zoo.* **39**, 1–48.

ALEXANDER F.E.S. & JACKSON R.M. (1955). Preparation of sections for study of soil micro-organisms. In: *Soil Zoology.* D.K. McE. Kevan (ed.). Butterworths, London. 433–441.

AUCAMP J.L. (1967). Efficiency of the grease film extraction technique in soil micro-arthropod surveys. In: *Progress in Soil Biology.* O. Graff & J.E. Satchell (eds.). Friedr. Vieweg, Braunschweig, North Holland Publishing Co., Amsterdam. 515–524.

AUCAMP J.L., LOOTS G.C. & RYKE P.A.J. (1964). Notes on the efficiency of funnel batteries for the extraction of soil micro-arthropods. *Tydskrif vir Natuurwetenskappe*, **4**, 105–126.

AUCAMP J.L. & RYKE P.A.J. (1964). A preliminary report on a grease film extraction method for soil micro-arthropods. *Pedobiologia*, **4**, 77–79.

BERG (1951). On the respiration of some mollusc from running and stagnant water. *Ann. biol., Paris,* **27**, 561.

BERTHET P. (1960). La mesure écologique de la température par détermination de la vitesse d'inversion du saccharose. *Vegetatio*, **9**, 197–207.

BERTHET P. (1963). Mesure de la consommation d'oxygène des Oribatides (Acariens) de la litière des forêts. In: *Soil Organisms.* J. Doeksen & J. van der Drift (eds.). North Holland Publishing Co., Amsterdam. 18–31.

BERTHET P. (1964). L'activité des Oribates d'une chênaie. *Mém. Inst. Roy. Sc. Nat. Belg.*, No. 151. 151 pp.

BERTHET P. (1967). The metabolic activity of Oribatid mites (Acarina) in different forest floors. In: *Secondary Productivity of Terrestrial Ecosystems.* Panstwowe Wydawnietwo Navkowe, Warsawa. 709–725.

BERTHET P & GÉRARD G. (1965). A statistical study of microdistribution of Oribatei (Acari). Part I. The distribution pattern. *Oïkos*, **16**, 214–227.

BERTHET P. & GÉRARD G. (1970). Note sur l'estimation de la densité de populations édaphiques. *Methods of Study in Soil Ecology.* UNESCO, Paris. 189–193.

BLOCK W. (1966). Some characteristics of the Macfadyen high gradient extractor for soil micro-arthropods. *Oïkos*, **17**, 1–9.

BLOCK W. (1967). Recovery of mites from peat and mineral soils using a new flotation method. *J. anim. ecol.* **36**, 323–327.

BOCOCK K.L. & HEATH J. (1967). Feeding activity of the millipede *Glomeris marginata* in relation to its vertical distribution in the soil. In: *Progress in Soil Biology.* O. Graff & J.E. Satchell (eds.). Friedr. Vieweg, Braunschweig. 233–240.

CANCELA DA FONSECA J.P. (1966). L'outil statistique en biologie du sol. I. Distributions de fréquences et tests de signification. *Rev. Ecol. Biol. Sol.* **2**, 299–332.

COHEN M. (1955). Soil sampling in the National Agricultural Advisory Service. In: *Soil Zoology.* D.K. McE. Kevan (ed.). Butterworths, London. 347–350.

COILE T.S. (1936). Soil samplers. *Soil Sci.* **42**, 139–142.

CROSSLEY D.A. (1966). Radioisotope measurement of food consumption by a leaf beetle species. *Ecology*, **47**, 1–8.

DAVIS J.J. & FOSTER R.F. (1958). Bioaccumulation of radioisotopes through aquatic food chains. *Ecology*, **39**, 530–535.

DEBAUCHE H.R. (1962). The structural analysis of animal communities of the soil. In: *Progress in Soil Zoology*. P.W. Murphy (ed.). Butterworths, London. 10–25.

DRIFT J. VAN DER (1950). Analysis of the animal community in a beech forest floor. *Tijdsch. Ent.* **94**, 94–168.

ELIASSEN E. (1953). Energy metabolism of *Artemia salina* in relation to body size, seasonal rhythms and different salinities. *Univ. Bergen Arbok, Naturvitenskap. Rekke*, **11**, 1–17. (cité par Florkin, 1960).

ENGELMANN M.D. (1961). The role of soil arthropods in the energetics of an old field community. *Ecol. Monogr.* **31**, 221–238.

ENGELMANN M.D. (1966). Energetics, terrestrial field studies and animal productivity. In: *Advances in Ecological Research*. J.B. Cragg (ed.). Academic Press, London and New York. 73–115.

FLORKIN M. (1960). Ecology and metabolism. In: *The Physiology of Crustacea*. T.H. Waterman (ed.). Academic Press, London and New York. **1**, 395–410.

FOURCHE J. (1963). Un respiromètre électrolytique pour l'étude des pupes isolées de Drosophile. *Bull. Biol. de France et de Belgique*, **98**, 475–489.

FOURCHE J. (1967). La respiration chez *Drosophila melanogaster* au cours de la métamorphose; influence de la pupaison, de la mue nymphale et de l'émergence. *J. Insect. Physiol.* **13**, 1269–1277.

GÉRARD G. (1967). Etude de la répartition spatiale de quelques populations d'Oribates (*Acarina: Oribatei*). In: *Progress in Soil Biology*. O. Graff & J.E. Satchell (eds.). Friedr. Vieweg, Braunschweig. 559–568.

GIFFORD D.R. (1967). An attempt to use ^{14}C as a tracer in a Scots pine (*Pinus silvestris* L.) litter decomposition study. In: *Secondary Productivity of Terrestrial Ecosystems*. Panstowe Wydawnietwo Nankowe, Warszawa. 687–693.

GROMADSKA M. (1949). The influence of constant and alternating temperatures upon CO_2 production of some insect chrysalide. *Studia Societaris Scientiarum Torunensis*, **2**, 1–28.

GROMADSKA M. (1966). Changes of respiratory metabolism of some land snails in different thermic conditions. *Zes zyty Naukowe Uniwersitetu Mikolaja Kopernika w Toruniu*, **9**, 101–107.

GROMADSKA M. & PRZYBYLAKA (1960). The influence of constant and alternating temperatures upon the respiratory metabolism of the snail *Arianta arbustorum* L. *Ecologia Polska*, serie A, **8**, 315–324.

GÖSSWALD K. (1963). Tracer experiments on food exchange in ants and termites. In: *Radiation and Radioisotopes Applied to Insects of Agricultural Importance*. Int. Atomic Energy Agency, Vienna. 25–42.

HALD A. (1952). *Statistical Theory with Engineering Applications*. John Wiley, London. 783 pp.

HARRIS M., HORVITZ D.G. & MOOD A.M. (1948). On the determination of sample sizes in designing experiments. *J. Amer. Stat. Assoc.* **43**, 391–402.

HARTENSTEIN R. (1961). On the distribution of forest soil microarthropods and their fit to 'contagious distribution function'. *Ecol.* **42**, 190–194.

HEALEY I.N. (1967). The population metabolism of *Onychiurus procampatus* Gisin (Collembola). In: *Progress in Soil Biology*. O. Graff & J.E. Satchell (eds.). Friedr. Vieweg, Braunschweig, North Holland Publishing Co., Amsterdam. 127–137.

HEALY M.J.R. (1962). Some basic statistical techniques in soil zoology. In: *Progress in Soil Zoology*. P.W. Murphy (ed.). Butterworths, London. 3–9.

HOLTER H. (1943). Technique of the cartesian diver. *C.R. Lab. Carlsberg, Ser. Chim.* **24**, 399–478.

HUBBEL S.P., SIKORA A. & PARIS O.H. (1965). Radiotracer, gravimetric and colorimetric studies of ingestion and assimilation rates of an isopod. *Health Physics*, **11**, 1485–1501.

KEMPSON D., LLOYD M. & GHELARDI R. (1963). A new extractor for woodland litter. *Pedobiologia*, **3**, 1–21.

KROG J. (1954). The influence of seasonal environmental changes upon the metabolism, lethal temperature and rate of heart beat of *Gammarus linnaeus* (Smith) taken from an Alaskan lake. *Biol. Bull.* **107**, 397–410.

LEBRUN PH. (1965). Contribution à l'étude écologique des Oribates de la litière dans une forêt de Moyenne-Belgique. *Mém. Inst. Roy. Sc. Nat. Belg.* No. 153, 96 pp.

LEBRUN PH. (1968). Ecologie et biocénotique de quelques peuplements d'Arthropodes édaphiques. Thèse de Doctorat. Univ. de Louvain. 462 pp., inédit.

LINDERSTRÖM-LANG K. (1943). On the theory of the cartesian diver micro-respirometer. *C.R. Lab. Carlsberg. Ser. Chim.* **24**, 333–398.

LØVLIE R. & ZEUTHEN E. (1962). The gradient diver. A recording instrument for gasometric micro analysis. *C.R. Lab. Carlsberg*, **32**, 513–534.

MACFADYEN A. (1952). The small arthropods of a Molinia fen at Cothill. *J. Anim. Ecol.* **21**, 87–117.

MACFADYEN A. (1953). Notes on methods for the extraction of small soil arthropods. *J. Anim. Ecol.* **22**, 65–77.

MACFADYEN A. (1961). Improved funnel-type extractors for soil arthropods. *J. Anim. Ecol.* **30**, 171–184.

MACFADYEN A. (1961). A new system for continuous respirometry of small air breathing invertebrates in near-natural conditions. *J. exp. Biol.* **38**, 323–341.

MACFADYEN A. (1962). Soil arthropod sampling. In: *Advances in Ecological Research*. J.B. Gragg (ed.). Academic Press, London. **I.** 1–34.

MACFADYEN A. (1966). Les méthodes d'étude de la productivité des invertébrés dans les écosystèmes terrestres. *La Terre et la Vie*, **4**, 361–392.

MARPLES T.G. (1966). A radionuclide tracer study of arthropod food chains in a *Spartina* salt marsh ecosystem. *Ecology*, **47**, 270–277.

MCMAHON E.A. (1963). A study of termite feeding relationships, using radioisotopes. *Ann. Ent. Soc. Amer.* **56**, 74–82.

MURPHY P.W. (1956). Soil fauna investigations. *R. on Forest Research*, **114**, 3.

MURPHY P.W. (1962). Sample preparation for funnel extraction and routine methods for handling the catch. In: *Progress in Soil Zoology*. P.W. Murphy (ed.). Butterworths, London. 189–198.

NIELSEN C.O. (1949). Studies on the soil microfauna. II. The soil inhabiting Nematodes. *Natura Jutlandica*. 131 pp.

NIELSEN C.O. (1961). Respiratory metabolism of some populations of Enchytraeid worms and freeliving Nematodes. *Oïkos*, **12**, 17–35.

O'CONNOR F.B. (1957). An ecological study of the Enchytraeid worm population from a coniferous forest soil. *Oïkos*, **8**, 161–169.

ODUM E.P., CONNELL C.E. & DAVENPORT L.B. (1962). Population energy flow of three primary consumer components of old-field ecosystems. *Ecology*, **43**, 88–96.

ODUM E.P. & KUENZER E.J. (1963). Experimental isolation of food chains in an old field ecosystem with the use of phophorus-32. In: *Radioecology*. V. Schultz & A.W. Klement, Jr. (eds.). Reinhold Corp., New York. 113–120.

ODUM E.P. & SMALLEY E.A. (1959). Comparison of population energy flow of herbivorous and a deposit-feeding invertebrate in a salt marsh ecosystem. *Proc. Nat. Acad. Sc. Wash.* **45**, 617–622.

PARIS O.H. & SIKORA A. (1965). Radiotracer demonstration of isopod herbivory. *Ecology*, **46**, 729–734.

PARIS O.H. & SIKORA A. (1967). Radiotracer analysis of the trophic dynamics of natural Isopod populations. In: *Secondary Productivity of Terrestrial Ecosystems*. Panstowe Wydawnietwo Nankowe, Warszawa. 741–771.

PENDELTON R.C. & GRUNDMANN A.W. (1954). Use of phosphorus-32 in tracing some insect-plant relationships in the thistle, *Cirsium undulatum*. *Ecology*, **35**, 187–191.

PHILLIPSON J. (1963). The use of respiratory data in estimating animal respiratory metabolism with particular reference to *Leiobunum rotundum* (Latr) (Phalangida). *Oïkos*, **14**, 212–223.

PHILLIPSON J. (1966). *Ecological Energetics*. Edw. Arnold, London, 57 pp.

PHILLIPSON J. (1967). Studies on the bioenergetics of woodland Diplopoda. In: *Secondary Productivity of Terrestrial Ecosystems*. Panstowe Wydawnietwo Nankowe, Warszawa. 679–685.

RAO K.P. & BULLOCK H. (1954). Q_{10} as a function of size and habitat temperature in poikilotherms. *Amer. Naturalist*, **88**, 33–44.

RAPOPORT E.H. (1966). Comentarios sobre la diataxis de algunos animales del suelo, con especial referencia a su distribucion espacial. In: *Progreses en Biologia del Suelo*. E.H. Rapoport (ed.). Centro de cooperacion cientifica de la Unesco para America Latina, Montevideo. 283–297.

REICHLE D.E. & CROSSLEY D.A. (1965). Radiocesium dispersion in a cryptozoan food web. *Health Physics*, **11**, 1375–1384.

REICHLE D.E. & CROSSLEY D.A. (1967a). Investigations on heterotrophic productivity in forest insect communities. In: *Secondary Productivity of Terrestrial Ecosystems*. Panstowe Wydawnietwo Nankowe, Warszawa. 563–587.

REICHLE D.E. & CROSSLEY D.A. (1967b). Trophic level concentrations of Cesium-137, Sodium and Potassium in Forest Arthropods. *Symposium on Radioecology*, D.J. Nelson and F.C. Evans (eds.). **Ann Arbor.** 678–686.

RODRIGUEZ J.G., PITAM SINGH, SEAY T.N. & WALLING M.V. (1967). Ingestion in the two spotted spider mite *Tetranychus urticae* Koch as influenced by wavelength of light. *J. Insect Physiol.* **13**, 925–932.

SMALLEY A.E. (1960). Energy flow of a salt marsh grass-hopper population. *Ecology*, **41**, 672–677.

Soo Hoo C.F. & Fraenkel G. (1966). The consumption, digestion and utilization of food plants by a polyphagous insect. *Prodenia eridania. J. Insect. Physiol.* **12**, 711–730.

Stein C.M. (1945). A two sample test for a linear hypothesis whose power is independent of the variances. *Ann. Math. Stat.* **16**, 243–258.

Teal J.M. (1962). Energy flow in the salt marsh ecosystem of Georgia. *Ecology*, **43**, 614–624.

Törne E. von (1962). A cylindrical tool for sampling manure and compost. In: *Progress in Soil Zoology.* P.W. Murphy (ed.). Butterworths, London. 240–242.

Vannier G. & Cancela da Fonesca J.P. (1966). L'échantillonnage de la microfaune du sol. *La Terre et la Vie*, **1**, 77–104.

Vannier G. & Vidal P. (1964). Sonde pédologique pour l'échantillonnage des microarthropodes. *Rev. Ecol. Biol. Sol.* **2**, 333–336.

Vernberg F.J. (1952). The oxygen consumption of two species of salamanders at different seasons of the year. *Physiol. Zool.* **25**, 243–249.

Wiegert R.G. (1964). Population energetics of Meadow Spittlebugs (*Philaenus spumarius* L.) as affected by migration and habitat. *Ecol. Monographs*, **34**, 217–241.

Wiegert R.G. (1965). Energy dynamics of the grasshopper populations in old field and alfalfa field ecosystems. *Oikos*, **16**, 161–176.

Williams E.C. & Reichle D.E. (1968). Radioactive tracers in the study of energy turnover by a grazing insect (*Chrysochus auratus* Fab.). *Oikos*, **19**, 10–18.

Wood T.G. (1965). Comparison of a funnel and a flotation method for extracting Acari and Collembola from moorland soil. *Pedobiologia*, **5**, 131–139.

Zeuthen E. (1947). Body size and metabolic rate in animal kingdom. *C.R. Lab. Carlsberg, ser. Chim.* **26**, 17–161.

Zeuthen E. (1950). Cartesian diver respirometer. *Biol. Bull.* **98**, 139–143.

Zeuthen E. (1953). Oxygen uptake as related to body size in organisms. *Quart. Rev. Biol.* **28**, 1–12.

Zinkler D. (1966). Vergleichende Untersuchungen zur Atmungsphysiologie von Collembolen und anderen Bodenkleinarthropoden. *Z. vergl. Physiol.* **52**, 99–144.

Apterygotes, Pauropods and Symphylans

I.N. HEALEY

Introduction

This chapter is concerned with methods for the study of production and energy flow in populations of Collembola, Diplura, Thysanura, Protura, Symphyla and Pauropoda. I understand these to be the processes involved in estimating the parameters of the energy budget equation as applied to single-species populations, trophic levels or communities. At the population level this is, in the terminology of Slobodkin (1962):

$$I = R + Y$$

where I = calories assimilated as food by the population, regarded as equivalent to total energy flow through the population,

R = heat loss or cost of maintenance metabolism, usually estimated by respiration, and measured in calories

and Y = production or yield from the population of potential energy in the form of tissues and excretory products, also measured in calories.

No further consideration will be given to the concepts and theory of energy flow except insofar as these relate to these arthropod groups. We still have little data on energy flow and production in these groups, but where relevant figures are available, these are quoted briefly to illustrate the level of sensitivity required of the techniques used, and to indicate the type of results to be expected.

Only one of these groups, the Collembola, can be regarded as generally prominent components of the soil fauna. In some soils the Collembola may be the most numerous of all arthropods, although generally they are exceeded in numerical abundance by the Acari, which have greater ecological diversity. (Raw (1967) reviews the comparative abundance of the arthropod groups in different soils and geographical regions). The other arthropod groups discussed here are generally considered to be rather rare animals, although, as shown, for instance, by Edwards (1958) for the Symphyla, when suitable extraction techniques are used they can often be shown to occur in large numbers. All of these forms are liable to show periods of relatively great abundance, usually for unknown reasons. [In October and November of 1964, for instance, a species of *Acerentomon* (Protura) was more abundant in a Kentish beechwood than any of 17 species of Collembola, of whom all but three were usually more common (Healey, unpublished)]. The rarity of some, especially the Symphyla and Pauropoda, may often be more apparent than real, due to the seasonal migrations which they make into deeper layers of the soil.

Table I summarizes the data available on population densities and biomass. There is a fair range of studies of Collembola populations in temperate grasslands of different types, but very few of woodland populations and no detailed systematic study from any tropical habitat. In the other groups there has been little detailed work of any kind.

Although the Collembola have a high population density they have, due to their small size, only very low biomass in comparison with that of earthworms, enchytraeids and nematodes, all of which have biomasses of many grams per m². Even the oribatid mites of deciduous woodland studied by Berthet (1964) have a biomass of 5·4 g/m² (adults only), about ten times the annual mean biomass for Collembola populations (though, of course, oribatids contain much metabolically inactive material). It is unlikely that the Collembola are of any direct quantitative significance in soil communities, either in terms of total energy flow or of exploitable biomass (Healey, 1967a). Quantitative significance is not, however, the only justification for the application of energy flow concepts to the study of populations. The measurement of productivity is a powerful tool for examining the structure and function of populations and communities and their relations with their environment. In the belief that this approach is profitable in any population, I have summarized in Table II the extent to which these six arthropod groups are suitable for studies of production and energy flow.

Methods for population and biomass estimation

(1) Sampling techniques

There have been several fairly recent reviews of the techniques and principles of extracting small arthropods from soil (Macfadyen, 1962; Murphy, 1962; Vannier and Cancela da Fonseca, 1966). Studies on the efficiency of these techniques for the various groups are in progress within the IBP and will be reported on by other workers (see Chapter 10). A detailed account of these is therefore unnecessary.

TABLE I. Estimates of population density and biomass in some soil microarthropods. Estimates marked with an asterisk are annual means derived from long-term studies and the rest are the results of occasional samplings.

Some of these estimates, those of van der Drift, Ryke and Loots, and Ostdiek, for instance, were made with probably rather inefficient extraction techniques.

Group	Author		Population density thousands/m²	Biomass Mg/m²
Collembola	Macfadyen (1952)	*Molinia* fen	25·0*	295*
		Deschampsia fen	24·0*	260*
		Juncus fen	7·2*	60*
	Hale (1966)	Limestone grassland	53·0*	330*
		Alluvial grassland	44·0*	370*
		Juncus grassland	21·0*	150*
		Calluna litter	35·0*	290*
	Dillon and Gibson (1962)	Pasture	33·0*	—
	Belfield (1956)	W. African pasture	25·0	—
	Milne (1962)	*Pteridium* moorland	16·6*	—
	Healey (1967a)	*Pteridium* moorland	42·0*	350*
	van der Drift (1951)	Beechwood	0·7	—
	Poole (1961)	Pinewood	40·0*	—
	Ostdiek (1961)	Deciduous woodland	9·0*	—
		Pinewood	24·5*	—
		Grassland	0·3*	—
	Ryke and Loots (1967)	African forest	5·0	—
Diplura				
Japygidae	Belfield (1956)	W. African pasture	1·5	—
Campodeidae	Salt *et al.* (1948)	Pasture	6·6	—
	Prater (unpub.)	Deciduous woodland	0·2–0·6	20–50
Thysanura	—	—	—	—
Protura	Salt *et al.* (1948)	Pasture	1·4	—
	Raw (1956)	Pasture	7·0	—
	Tuxen (1949)	Beechwood	6·5*	—
Symphyla	Salt *et al.* (1948)	Pasture	3·9	
	Edwards (1958)	Cultivated soil	0·6–19·0	—
	Belfield (1956)	W. African pasture	10·8	—
Pauropoda	Salt *et al.* (1948)	Pasture	0·6	—
	Starling (1944)	Pinewood on sand	0·5*	—
		Oakwood on clay	0·3*	—
	Belfield (1956)	W. African pasture	6·2	—

TABLE II. Amenability of 6 arthropod groups to the various components of productivity and energy flow studies, with the techniques available at present.

Group	1	2	3	4	5	6	7	8	9	10
Collembola	5–25	Difficult but possible	Yes	Yes, with qualifications	No	Many species	Some species	Most species	Difficult	Possible if 9 successful
Diplura										
Japygidae	0–2?	Yes	No	Yes	Possibly	Yes	Yes	Yes	Yes	Yes
Campodeidae	1–2	Yes	Yes	Yes	Doubtful	Yes	Yes	Yes	Difficult	Yes if 9 successful
Thysanura	0–1	Yes	No	Yes	Yes	Yes	Yes	Some species	Yes	Some species
Protura	0–3	Difficult but possible	Yes	Yes	Probably	Yes	Doubtful	Doubtful	?	?
Symphyla	0–2	Yes	Yes	Yes	Doubtful	Yes	Probably	Yes	Yes	Yes
Pauropoda	0–5?	Difficult	Yes	Yes	Probably	?	Doubtful	Doubtful	?	?

1. Number of species likely to be found in a particular habitat.
2. Identification.
3. Quantitative sampling and population estimation.
4. Age structure analysis.
5. Cohort analysis.
6. Weighing and biomass estimation.
7. The study of metabolism of individuals.
8. Culturing.
9. Identification of diet.
10. Quantitative study of feeding.

Extraction techniques are conventionally classified into two types, the so-called 'dynamic' methods which use the behavioural responses of the arthropods to externally-applied stimuli of heat and desiccation to drive them from the soil sample, and mechanical methods in which the arthropods are separated from the soil and plant debris, usually by using a combination of their different densities and wetting properties. The efficiency of dynamic methods depends on the ability of the arthropods to perceive such stimuli and to respond to them by movement, and is therefore a physiological problem whilst the efficiency of mechanical methods is only a technical problem. One would expect the latter to be more readily soluble and mechanical methods to have wider applicability. But in practice no mechanical method of separating arthropods from soil has been devised that is both efficient and applicable to a wide range of soils or arthropod groups. All are time-consuming and difficult to use without extensive laboratory facilities, and none permit the collection of animals alive.

The dynamic method currently most widely used for extraction of Collembola is the 'high gradient' canister extractor, a modification of the Tullgren funnel, designed by Macfadyen (Macfadyen, 1961; Block, 1966). The soil sample, held in a ring of heat-resistant plastic, is inverted (i.e. placed vegetation-downwards) into the top of a metal canister on a rust-proof grid, which for the extraction of Apterygota should have holes about 3 mm diameter (i.e. around 10 holes per cm²). The sample holder is then fitted through a heat-proof baffle of plastic or asbestos so that the canister projects into a bath of circulating cold water. The sample is then gently heated from above by a light bulb or resistance-wire heater. It is simultaneously cooled and kept moist from below by the cold damp air within the canister. A steep gradient of temperature and desiccation is thus set up within the sample. The arthropods are presumed to pass along this gradient away from the heat source and collect at the cool humid base of the sample. Heating is gradually increased in the course of the extraction and eventually the animals leave the sample, passing through the grid, and drop into a little fixative (usually a 40% saturated aqueous solution of picric acid).

Full details of the design and construction of canister extractors are given by the two authors cited above. Macfadyen suggests that during the greater part of the extraction the temperature of the heated surface of the sample should be between 30° and 40°C, and of the cooled surface 15°C, giving a gradient of 15—20°C. The heating should be briefly increased for the last few hours of extraction to dry out the base of the sample. As an extremely rough guide, for an apparatus designed to take, say, 36 small samples of 3–5 cm width and 3 cm depth an adequate heating regime can be given by a 60W light bulb controlled by a variable transformer run at 100 volts for the greater part of the extraction, and with cooling water at about 8–10°C. Extraction of Collembola might take three to four days.

The strength and durability of the temperature gradient within the sample depends to a considerable extent, as Macfadyen (1968) has recently shown, on the thermal properties of the particular soil. The strongest gradients are given by highly compact organic soils such as those of pasture grassland, and the weakest by non-organic sandy or clay soils on the one hand and leaf litter on the other. Some soils, particularly woodland ones, are therefore likely to vary in their ability to sustain a gradient at different times of year. Consequently it is important in the design of canister extractors that everything possible must be done to maximize gradient formation. Any lateral heating of the sample must be avoided by the use of heat-resistant plastic sample holders, by really heat-proof baffles and by having the sample project as little above the baffle as possible. Canisters of metal rather than plastic should always be used, to increase the conduction of cold from the water bath to the base of the sample. If a good gradient is not obtained with the heated surface of the sample at 30—40°C it is preferable to increase the rate of cooling by dropping the temperature of the water bath to 3 or 4°C, rather than to increase the heating further. Finally, it is important that the heating of samples within the extractor should be uniform; if there is any doubt whether the air currents above the samples give adequate stirring of the air a small fan should be installed.

The collecting fluid at the base of the canister presents a large surface for evaporation to the base of the sample and so it is important that an odourless collecting fluid should be used in case it should inhibit animals leaving the sample. Animals may be collected into water, or if it is desired to kill them to prevent damage by predators, an aqueous solution of picric acid may be used. (If the Collembola are wanted alive for experimental purposes, it is better to collect them on to damp plaster of paris rather than water, to prevent bloating).

The absolute efficiency of the canister extractor is hard to evaluate. Macfadyen (1955) showed by direct comparison with hand-sorted samples that some relatively primitive Tullgren funnels extracted around 35% of the Collembola fauna. When these funnels were used in a direct comparison with canister extractors, the latter were found to extract about 4 times as many

Collembola, and Macfadyen concluded that, at least in favourable conditions and soils, extraction could be nearly perfect. This is supported by the fact that it is usually very difficult to find any Collembola remains in soil samples after extraction, either by hand-sorting (the efficiency of which must of course be very low) or by mechanical extraction. Block (1966) added known numbers of Acari to sterile soil samples and extracted them in a high-gradient canister extractor; he found that only 76% were recovered. This may, of course, be because the mites were disorientated by their treatment and for such an experiment to be conclusive it would be necessary to use control batches of mites kept in culture to check natural mortality rates. It may in any case be that the more highly-mobile Collembola would behave very differently in such an experiment, but as far as I know no strictly comparable test has been made with the Collembola.

Mechanical methods of extraction of arthropods from soil have reached their greatest development in the flotation techniques of Raw (1955) and Hale (1964). In Raw's technique the sample, which should be about 50 cm^3, is subjected to a series of processes including colloid dispersion in sodium hexametaphosphate, freezing and sieving, which break open the soil particles. It is then agitated in a magnesium sulphate solution of specific gravity 1·2 in which all organic material separates from the soil. After washing to remove the $MgSO_4$ the arthropods are separated from the plant material by shaking in a benzene/water mixture; the arthropods are wetted by benzene and become suspended above the benzene/water interface, whilst the plant matter is not wetted and remains in the water layer. The suspension is then cooled to about 4°C when the arthropods may be removed in the frozen plug of benzene.

The efficiency of this technique is again hard to evaluate. It was designed for use on arable and grassland soils containing relatively small amounts of organic matter. The Collembola probably extract better than most other arthropod groups because most species are not wettable, and it is likely that in the hands of an experienced operator on suitable soils this technique may give nearly perfect extraction of Collembola. But it has three major disadvantages: (i) it is extremely laborious and time-consuming; (ii) it requires fairly extensive laboratory facilities, and (iii) the animals extracted are frequently too badly damaged for taxonomy or age structure analysis.

Hale's flotation technique (1964) was designed specifically for Collembola in peat soils. The sample is broken down by passage through a series of sieves, and is then placed in a vacuum flask, either in water if Collembola are to be extracted, or in magnesium sulphate solution of 1·2 S.G. for other arthropod groups. The sample is then caused to boil by pressure reduction so that the peat particles are dispersed. Air is now bubbled through the sample and the arthropods float to the surface where they may be decanted off by a side-arm. Hale compared the efficiency of this technique with that of a high gradient canister extractor, using peat samples. He obtained consistently slightly higher numbers of Collembola by the flotation technique, probably due to the inclusion of recently dead but undecomposed specimens, but for total Collembola the difference was not statistically significant. A rather large species, *Onychiurus latus*, was significantly better extracted by flotation, but this may have been because it was unable to pass through the grid in the canister extractor. Another species, *Isotoma viridis*, which is large and very active, showed similar results; this and other surface-active species may escape during the setting up and running of Tullgren extractions, but are rarely quantitatively important. In general, it was encouraging that the two techniques gave such similar results for the number of Collembola. [Block (1967), however, has compared the efficiency of Hale's technique with that of a canister extractor for the extraction of mites from peat and mineral soils. In contrast to the results for Collembola, he found that the canister extractor was much more efficient than the flotation technique; it extracted twice as many mites from peat, and $3\frac{1}{4}$ times as many from mineral soils.] The chief disadvantages of the technique are the cumbersome apparatus and extensive laboratory facilities required.

Flotation and Tullgren techniques are not equally suitable for use on all types of soils. Raw's technique was developed for largely mineral soils. It is probably less efficient on woodland and other highly organic soils in which the animals are not readily separated from tangled plant and fungal matter, or are trapped between pieces of leaf litter. [Work by Satchell and Nelson (1962), however, who used this technique to extract Acari from woodland soils, suggested that variability in size, body form and surface properties between species might be more important problems].

Wood (1965) compared Raw's technique with a rather primitive Tullgren funnel for the extraction of Collembola from four soils, a mull-like rendzina and three different brown earths. In most cases he obtained higher numbers with the funnels, but two species, *Entomobrya albocincta* and *Sminthurinus niger*, were found in higher numbers by flotation. An important point, though not fully analysed by Wood, was that for some species the funnels seemed less efficient on some soils, especially the gleyed podsols.

In general, mineral and clay soils show a tendency to compaction and shrinking in Tullgren funnels which may hinder the animals leaving the samples. Milne (1962) found this problem so serious that he used a high-gradient canister extractor for the litter and humus layers and Raw's technique for the mineral soil. If contraction of the sample occurs in the early stages of extraction it is usually a sign that the heating regime is too powerful. High-gradient Tullgren techniques are also not altogether suitable for litter-covered woodland soils. Samples must normally be inverted in the Tullgren method, because surface forms are unable to burrow through humus and mineral soil to leave the sample; but when woodland soils are inverted the larger litter fragments form a barrier to the soil forms. A further difficulty is that in coarse-grained litter animals may become trapped in pockets of moisture between pieces of litter (Macfadyen, 1955). [In temperate conditions the importance of these problems will, of course, vary with season.] The first problem can be alleviated by extracting the litter and soil portions of a sample separately. Kempson, Lloyd and Ghelardi (1963) have developed an extractor specifically for woodland litter, which employs the high-gradient principle but takes large samples (around 0·5 litres). Heating is by infra-red lamps, which have greater penetration than conventional heating techniques, and the temperature above the samples is controlled by a Simmerstat. A large surface area of collecting fluid is exposed only a few centimetres below the base of the sample, which maintains a very high humidity at this point. Extraction efficiency is probably better than 90% for Collembola (and probably comparable for Pauropoda). How this apparatus would adapt to the small samples necessary for accurate population estimation is not known.

To summarize this discussion of extraction techniques for Collembola: flotation techniques, whilst in theory preferable to behavioural techniques, cannot generally be recommended for use in productivity studies. They require good laboratory facilities, are extremely time-consuming and are not suitable for the large-scale replication needed for modern techniques of statistical analysis. High gradient Tullgren techniques, on the other hand, are capable of almost infinite replication and are flexible in design. Macfadyen (1962) for instance has designed a high-gradient canister apparatus for use under expedition conditions which employs as a heating element a metal plate suspended over the sample which is heated at the sides by paraffin or gas burners. In the conventionally-heated canister extractor a heating regime can be selected which is suited to the soil and species under investigation, and this is indeed essential. But there is no universal extraction technique that can be applied indiscriminately to all soils and for all species. High-gradient canister type extractors are probably at their least efficient on woodland soils. It must be borne in mind that we know almost nothing about the factors that cause Collembola to leave soil samples under Tullgren extraction, and nothing about the behaviour of Collembola in the gradients of temperature and humidity that develop in such samples. Every effort should be made to test the efficiency of the chosen technique for the soils and species under study.

The design of corers for taking soil samples is important. Corers should be designed so that the sample passes straight into the plastic ring in which it is to be extracted, thus reducing disturbance of the core to a minimum; a suitable general design is given by Macfadyen (1961). The compression, vertical as the corer cuts into the soil, and lateral as the soil passes into the sample holder, is probably a serious source of error in the technique, particularly when small samples are being taken. Hughes (1954), who has made the only critical study of this problem, showed that recovery of Collembola could be reduced by the use of only a very slightly blunt corer. Here again, the problem is most serious with woodland soils; leaf litter undergoes violent lateral shearing as well as vertical compression as the corer enters. No satisfactory corer for woodland soils has as yet been designed; it is possible that some ultra-fast-cutting, power-driven tool will be required.

With regard to the other arthropod groups under consideration, rather little information is available on which to base a choice of extraction techniques. Raw (1956) found high populations of Protura in grassland (Table I) using his flotation technique, but Tuxen (1949) in a more comprehensive study in a beechwood, found high population densities using a primitive Tullgren apparatus, and still (Tuxen, 1964) recommends this method. Protura have poor locomotory powers and high gradient funnels cannot be ideal for extraction but, because of their spiny bodies and limbs, they extract badly from woodland soils by flotation, and in a condition too poor for age structure analysis. For Diplura, Pagés (1967) considers no extraction technique suitable for *Dipljapyx* and feels that population counts are best made by hand collecting. Prater, in our laboratory, has found simple, uncooled Tullgren funnels taking large samples, and heated very slowly (extraction takes 12–14 days) suitable for *Campodea*; flotation was unsuccessful. Edwards (1962) reviewed extraction techniques for Symphyla, and found that much higher population figures have been obtained by flotation, probably because Symphyla cannot move freely through soil; he feels, however, that Tullgren funnels may still be preferable for woodland litter soils. The single ecological study of the Pauropoda (Starling, 1944) utilized simple Tullgren funnels; high numbers have also been found by flotation techniques (Salt *et al.*, 1948; Belfield, 1956) and by the litter extractor of Kempson, Lloyd and Ghelardi (1963).

(2) Design of sampling surveys

Population estimates for soil microarthropods typically have very low accuracy. Co-efficients of variation for sets of sample data are frequently around 50–100% giving 95% confidence limits no better than up to ±50% of the mean. Whatever our accuracy in the estimation of biomass and metabolism may be, the accuracy, or lack of it, of our estimates of production and population metabolism cannot be greater than that of our population estimates. But the reproducibility of population estimates is in fact greater than these figures would suggest (Debauche, 1962), since much of the variance is due to the highly aggregated distribution shown by most soil microarthropods. Once this effect has been understood, and its intensity measured, much of the variation can be stabilized by subjecting data to appropriate transformation, and confidence limits for the mean reduced to a more reasonable ±10–20% of the mean. In these circumstances it is important that sampling surveys should be designed with great care, particularly with regard to size and number of sample units, so that the nature of the distribution of the population, and the transformation required to analyse it, can be clearly understood. The general principles of the design of sampling surveys are discussed by Macfadyen (1962), who quotes the relevant statistical authorities, and methods of measuring aggregation, and of deciding what transformations are required are dealt with by Debauche (1962), Taylor (1961) and Cancela da Fonesca (1966). Increasingly the estimation of productivity parameters from population data will become a matter of complex multivariate analysis, and there can be no substitute for specialist statistical advice at an early stage.

Collembolan populations are usually aggregated on two separate levels; areas superficially uniform in vegetation and soil type may contain regions of greatly varying population density (e.g. Milne, 1962) and within these regions the population consists of a series of aggregates of varying density and width. Hughes (1962) showed that in *Folsomia* the width of such aggregates varied from about 10 cms to 35 cms, and that density within them varied widely. The origin of clumping in Collembola and in other microarthropods, is still far from clear (e.g. Poole, 1962; Berthet and Gerard, 1965).

For efficient population estimation it is important that the size of sample unit used should be small in comparison with the scale of clumping shown by the population. In general, it is statistically preferable to take a large number of rather small samples than fewer big ones. Preliminary sampling should be carried out using a number of different sizes of sampling unit, from which the optimum, that giving the lowest ratio of standard error of the mean to the mean, should be selected. Unfortunately the degree of aggregation shown by the population will vary with density at different times of the year (Healey, 1965), so that a given size of sample unit will not give equal accuracy at all times. Generally a unit of about 10 to 25 cm² area is best for Collembola; a 25 cm² unit gives a manageable number of Collembola of about 50–100. Little is known about the size of sample required for the other arthropod groups. Where density overall is rather low an inconveniently large sample unit may be necessary to avoid a large number of nil counts. We have found a sample unit about 300 cm² in area necessary for *Campodea* in woodland; a much smaller unit would suffice in grassland.

The number of sample units that must be taken for a given degree of accuracy of population estimation with a given size of sample unit can be calculated from the preliminary sampling data by the usual statistical procedures (Macfadyen, 1962, and standard textbooks). This calculation should be carried out with the data subjected to approximately the same transformation that will be used for final analysis of the data. [All these calculations must, of course, be tempered by a knowledge of the time, money and facilities available for the investigation]. In taking the samples it is probably better to distribute them over the study area by a stratified random scheme to ensure even coverage, rather than to rely on a completely random distribution.

The frequency of sampling must be related to the rate of age structure changes shown by the population, an understanding of which is necessary for estimates of production. The duration of life history in different species of Collembola varies from a few weeks in the ecologically-important *Folsomia* and in many Sminthurids to six months to a year in many onychiurids and entomobyrids, and so it is likely that some species have at least two or three generations per year even in temperate regions. Samplings at only monthly intervals, as is conventional, will frequently mask important changes in age structure, as I have found (Healey, 1967b), even in the rather slow-growing *Onychiurus procampatus* Gisin. Thus, even in studies in temperate regions sampling at fortnightly intervals is to be recommended and is (presumably) even more essential in tropical studies. Sample replicates tend to be more uniform in the age distribution of the animals they contain than in their total number, and so it may be possible to run a full-scale sample survey at monthly intervals in combination with a smaller number of samples taken for age structure analysis at more frequent intervals.

Protura probably have annual life histories (Tuxen, 1949, and personal communication) so that monthly sampling, at least for most of the year, would probably give an adequate picture of age structure changes. Symphyla probably have an annual cycle (Edwards, 1959), but show rather abrupt changes in numbers by monthly sampling and so rather frequent sampling is probably necessary.

The depth in the soil to which sampling is necessary can usually be readily estimated from a preliminary sampling survey and varies very widely in different soils. The possibility of seasonal, or in some cases daily, vertical migrations by the population should be borne in mind.

Estimation of age structure, biomass and secondary production

(A) Age structure

Production by a population may be estimated either directly by the computation of rates of absolute growth of individuals, or indirectly as yield to predators or decomposers from estimates of mortality. In both cases a knowledge of the significance of age structure changes in the population is necessary. There are two main problems with Collembolan populations: (i) it may be difficult to interpret age structure in terms of the conventional arthropod instar classes and (ii) continuous, or at least protracted, recruitment to the population may make definition of generations difficult.

(i) Some authors, for instance Agrell (1948) and Hale (1965b) [who also reviewed the number of instars found in different species of Collembola] have identified instars and defined size ranges for them, by breeding experiments in the laboratory. Hale found that the length of the head capsule is a good criterion of instar in *Hypogastrura, Onychiurus* and *Tullbergia*, and was able (Hale, 1966b) to use this to establish the age structure of field populations. However, as the work of Cassagnau (1961, and other papers) has emphasized, body form in the Collembola is extremely plastic in response to environmental factors. The number of pre-adult instars, their size range, and especially their duration, vary with feeding and seasonal conditions. As is well known, the Apterygota continue to moult after maturity, and if conditions are adequate these post-adult instars may show small size increments. Instar criteria established for a species on one habitat may not be applicable elsewhere (Raynal, 1964; Healey, 1967b). It may be possible to define instars on the basis of chaetotaxy (Hale, 1965b) but as the variation between individuals is very great this is probably best avoided by ecologists. In view of these difficulties, and the great amount of work required to define instar classes, it may be better in ecological work to classify the population into arbitrary size classes, such as the simple size/weight classes that I have used for *Onychiurus procampatus* (Healey, 1967b). These have definite limits and unlike instars there is no overlap between classes, and their width can be chosen to give any required degree of precision in the estimation of biomass and population metabolism. It may eventually be possible to apply the techniques for analysing polymodal frequency distributions (e.g. Taylor, 1965) that have been so successful in, for instance, millipedes (Blower and Gabbutt, 1964): personally, I have not yet been successful in doing so.

(ii) Even in univoltine species of Collembola recruitment to the population by hatching of eggs occurs generally over a protracted period (except in some surface living species), so that it is usual for most size classes to be present all through the year. Populations are not therefore generally susceptible to the conventional cohort analysis for the estimation of production. In the few cases in which curves of growth in average weight of a cohort have been calculated (e.g. by Agrell, 1948, on some Arctic forms which hatch suddenly at spring thaw) these have been of the usual S-shaped form. No survivorship curve for any Collembolan population has been established with certainty, although by making certain assumptions about growth rates I was able to construct an 'artificial' one for *O. procampatus* (Healey, 1967b). Richards, Waloff and Spradbery (1960) have described a method for cohort analysis and the estimation of mortality in populations in which recruitment is protracted. It is possible that this method could be applied to Collembolan populations, but it will be necessary to know rather more about the initial size of cohorts (i.e. the fecundity of the preceding generations) and about growth rates than we do. At the moment no methods for analysing the type of complex age structure shown by the Collembola are available.

The age classes of Protura can be identified by chaetotaxy and the number of abdominal segments, and since there is probably only one generation per year (Tuxen, 1949), age structure and cohort analysis should be possible when suitable population studies are made. The definition of both size classes and cohorts would seem to present no difficulties in the Machilidae (Thysanura) (Delaney, 1957), but no population study has been made. Age classification is difficult in *Campodea* (Diplura), but cohort analysis may, doubtfully, be possible. The Symphyla and Pauropoda can be divided into age classes on the basis of the number of abdominal segments and chaetotaxy, but since the Symphyla have long recruitment periods, cohort analysis might prove difficult.

(B) Biomass

Collembola vary in weight from about $0·1\mu$ in the hatchlings of many small species to over 1 mg in the adults of some large entomobryids. Eggs vary from about $0·1\mu$ to perhaps 5 or 6 μg. Size for size it is likely that the other groups under consideration will have comparable weights, except for acerentomid Protura which are more heavily sclerotised and are somewhat heavier. Electrobalances are now available that have sensitivity of 1 μg or better and can be used for rapid weighing of delicate animals of this order of size, in some cases under field conditions. Many Collembola are hard to weigh because of their activity and ability to jump. The larger species can be weighed in small gelatin capsules, which will only marginally decrease the sensitivity of weighing. But many small species (poduroids, young *Folsomia*, small sminthurids, pauropods) cannot be weighed individually because of these handling difficulties. In cases where these are ecologically important and the weight variation of individuals must be known, or where an electrobalance is not available, it seems that the only alternative is to "weigh" animals by chemical microanalysis. This has not been used for terrestrial microarthropods as far as I know, and it may not have the sensitivity required, but it has been used for marine plankton of comparable size by Zeuthen (1947) and for Enchytraeidae by Nielsen (1961).

Various factors, such as the collection of sample material directly into preservative, make it necessary for estimation of biomass of samples to be indirect. If instars can be identified average weights for individuals in each instar can be established (Hale, 1966a) and these applied to age structure analysis to give biomass estimates. There is great variation in size within instars, however (a 4th instar *O. procampatus* may vary in weight from 20 μg to over 50 μg), and this procedure must be handled carefully. An alternative is to establish a statistical relationship between weight and a body dimension that can be quickly measured on preserved sample material. Healey (1965, 1967b) found that head length gave the best relationship with live weight in

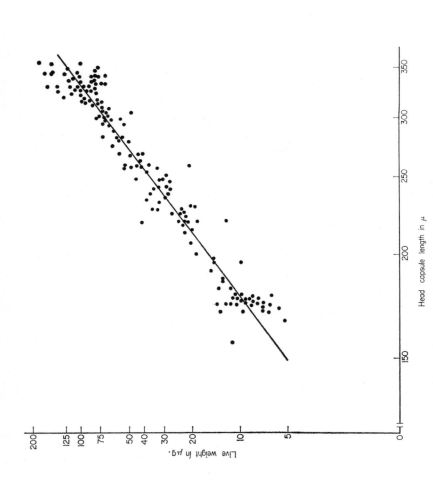

Figure 1. The relationship between head capsule length and live weight in *Onychiurus procampatus* Gisin.

O. procampatus, and established a regression (Fig. 1) from which the weight of field animals could be estimated.

Biomass of some Collembolan populations is given in Table I. These figures are rather uniform and are low compared with some earlier published estimates of Collembolan biomass (e.g. that of 5·0 g/m² for grassland by Macfadyen, 1963).

For some purposes, such as calculation of standing crop, the dry weight of the population must be known. Some problems may occur in the conversion of biomass data to dry weight. Healey (1965) found that *O. procampatus* from the field had a mean dry weight of 45·0±S.E. 0·8% of live weight. Figures of 20–35% are typical for insects and so these figures were surprisingly high, especially for a soft-bodied arthropod. Fat extraction showed these animals to have a high fat content of 54% of dry weight (24% of live weight). Subsequent work has shown that this situation is typical only at certain times of the year. *O. procampatus* follows a cycle of fat deposition with increasing age in which the highest fat contents are found in post-breeding, senile adults in late spring. Blackith and Goto (Blackith, 1962) found that the water content of Collembolan species varied in different habitats. The significance of these facts is not known, but care should clearly be exercised in converting biomass estimates to dry weights on the basis of what may be seasonally biassed estimates of water content. [Prater (personal communication) has found a dry weight of 23% of live weight to be applicable to *Campodea staphylinus* at all times of the year].

(C) Secondary production

If growth rates and survivorship curves for cohorts can be established it is fairly straightforward to combine these with biomass figures to obtain an estimate of secondary production by the population. The studies of Wiegert (1965) on grasshoppers and Saito (1965) on a woodlouse would form suitable models. Briefly, most secondary production can be calculated from the average weight of the cohort in a sampling period multiplied by the mortality for that period. Wiegert's paper discusses various aspects of this technique and some of the mathematical formulations which underlie it are dealt with by Neess and Dugdale (1959). Various other components of secondary production must be added, of which the most important are reproductive materials, which can be estimated from fecundity rates for the population (collembolan fecundity rates are reviewed by Hale, 1965a). Production by cast exuviae is important in many insects. In Collembola these are only 2–3% of body weight, but since some apterygotes moult up to 60 times in their lives this could be an important factor.

No comprehensive study of this kind has been made for an apterygote population. However, in *Onychiurus procampatus*, by making certain assumptions about mortality and growth rates I was able to suggest (Healey, 1967b) that production was of the order of a little over four times annual mean biomass, and to confirm roughly the magnitude of this figure by summing positive biomass increments. But this approach cannot be regarded as a substitute for a proper cohort analysis.

The low biomass of these forms means that only little material of each stage in the life history can be collected for bomb calorimetry. Only the instrument designed by Phillipson (1964), which is now commercially available, has the sensitivity required. It can be used for samples as small as 1 or 2 mg dry weight (about 30–40 adult *Onychiurus*, for instance). Work with this apparatus is in its early stages and it would be unwise to quote any figures for calorific content.

The estimation of population metabolism

Leaving aside the early work of Bornebusch and others, only two studies of the metabolism of Collembola have been made (Zinkler, 1966, and Healey, 1965, 1967a); of these the former covered 13 species, but was not applied to field populations, whilst the latter dealt with only one species but was used to estimate population metabolism. The metabolic rates of Collembola at ecologically-significant temperatures (0°C to 20°C) are roughly around 0·002 μl O_2/individual/hour (perhaps 1,000 μl O_2/g/hr) to 0·8 μl O_2/individual/hour (50 μl O_2/g/hr). They are roughly comparable with nematodes and enchytraeid worms of similar weight.

Equating the study of metabolism with the study of respiration (no technique for the estimation of direct heat dissipation by animals like these having been developed), very few techniques are sufficiently sensitive for the study of individual respiratory rates. Zinkler (1966) used a miniaturized Warburg apparatus whose chief modifications were the use of a very small reaction vessel of 0·8 ml capacity, a micromanometer immersed in the water bath and sensitive to volume changes equivalent to 0·05 μl of oxygen. Sensitivity was thus rather poor relative to the respiratory rates of many Collembola, and Zinkler was not able to work with animals of less than about 80 μg weight. The respiration of smaller species could only be estimated by leaving them in the apparatus without food for very long periods. For instance, the respiratory rate of a 4th or 5th instar *Onychiurus* weighing 50 μg at 15°C is about 0·023 μl O_2/hr, so that readings even at minimal sensitivity would be possible only every two hours. However, Zinkler was able to make a comprehensive study of respiration relative to temperature and live weight in several larger species, and his results will be of great value to ecologists. He also showed that the respiratory rate of 'active' individuals could be twice that of 'resting' individuals which raises the general problem of the applicability of laboratory results to field populations. But it would seem that this modified Warburg apparatus, which is expensive and not generally available, cannot have wide application in ecological studies of the Collembola.

Only one technique, the Cartesian diver microrespirometer, has the sensitivity required. This technique is generally applicable to respiratory rates below 0·1 μl O_2/hr and has sensitivity of at least 0·001 μl O_2. Detailed accounts of the technique are given by Glick (1961) and Zeuthen (1964) who give all necessary references. Briefly, the animal is enclosed in a small glass vessel, the diver, open at one end and floating in a liquid. As the animal respires and CO_2 is absorbed by alkali, the volume of the gas phase in the diver decreases and the buoyancy of the diver changes; these changes can be measured manometrically and used to estimate respiration. The so-called "stoppered" diver described by Zeuthen (1950) is now accepted as the easiest to operate in most circumstances. The technique is far from ideal for small terrestrial arthropods, since the inside of the diver is quite unlike the natural environment, and setting up the diver involves considerable handling of the animals but, as the comprehensive study of Berthet (1964) shows, results are surprisingly consistent. I used this technique for a study of *Onychiurus procampatus* and have obtained estimates of metabolic rates of a number of other species. It is satisfactory for species that lack a furca, but more active species are difficult to place in the diver, and no answer to this problem has been found.

The estimation of assimilation

No estimates of assimilation have been made for any apterygote population (beyond that based on many assumptions by Healey, 1967b) and undoubtedly the main obstacle to the study of feeding in these forms lies in establishing the natural diet (Poole, 1959; Sturm, 1959) which in all cases is poorly known. Much can be learnt from careful analysis of gut contents, but relative to experimental conditions, the sensory equipment of these forms is so poor that food selection experiments are often meaningless (Healey, 1965) and food preferences are probably best demonstrated by growth and fecundity rates. Where diet can be identified simple quantitative study of feeding is possible by weighing small samples of food before and after feeding. This technique has been applied to Symphyla by Edwards (1961) and with limited success to *O. procampatus* (Healey, in prep.).

Culture techniques are important in the study of feeding, as well as of growth and fecundity. The techniques of Edwards (1955) and Goto (1961) are recommended. Many species of Collembola have been successfully cultured and they may be fed on dried yeast, *Pleurococcus*, fungi (especially *Mortierellas* and *Mucors*) and, in some cases, well-rotted leaf litter. Species of Diplura and Symphyla have also been cultured. Protura will not normally survive long in culture, but may be kept for 2—3 months by adding onychiurid Collembola who will control the growth of fungi, which will otherwise trap the Protura (Healey, unpublished).

An apparatus which helps in the handling of the more active micro-arthropods is now available. This is the thermomodule which operates on the principle of the Peltier effect and is used to cool microtome blades. Small, thin-bottomed culture dishes can be placed on the module and their temperature quickly lowered until the animals become sufficiently sluggish to permit handling. This apparatus is commercially available and is also described by Vannier (1965).

Conclusions

No general conclusions on energy flow in these groups is possible beyond the obvious one that relative to other soil organisms it will be very small. Mac-fadyen (1963) suggested that the fauna accounted for only 20% of the total energy released in the soil and it is unlikely therefore that these forms account for more than 1% at most. But it is likely that their importance lies in the promotion, for instance by spore distribution (Talbot, 1952; Frankland, 1966), of the activities of other soil organisms. In any case, with their high numbers, simple development and activity all through the year, they make good subjects for studies of the seasonal pattern of energy flow.

I would suggest that there are four main areas in the study of energy flow and production in soil microarthropods in which our knowledge is noticeably deficient:

(i) The efficiency of extraction techniques and their applicability to particular soils and species.

(ii) The analysis of age structure, cohort identity and absolute growth rates in populations with continuous or protracted recruitment.

(iii) The application of laboratory estimates of metabolism to field populations.

(iv) The nature of the diet, and feeding behaviour and the quantitative study of feeding in general.

Acknowledgements

I am grateful for research support to the Agricultural Research Council and to the Central Research Fund of the University of London. I am also grateful to Professor A. Macfadyen for constant advice and encouragement.

References

AGRELL I. (1948). Studies on the postembryonic development of Collembola. *Ark. fur Zool.* **41**, 12, 1–35.

BELFIELD W. (1956). The arthropoda of the soil in a West African pasture. *J. Anim. Ecol.* **25** 275–287.

BERTHET P. (1964). L'activité des Oribatides (Acari: Oribatei) d'une chênaie. *Mém. Inst. Roy. Nat. Belg.* **152**, 1–152.

BERTHET P. & GÉRARD G. (1965). A statistical study of microdistribution of Oribatei (Acari). Part I. The distribution pattern. *Oïkos*, **16**, 214–227.

BLACKITH R.E. (1962). The handling of multiple measurements. In: *Progress in Soil Zoology*. P.W. Murphy (ed.). Butterworths, London. 37–42.

BLOCK W. (1966). Some characteristics of the Macfadyen high gradient extractor for soil microarthropods. *Oïkos*, **17**, 1–9.

BLOCK W. (1967). Recovery of mites from peat and mineral soils using a new flotation method. *J. Anim. Ecol.* **36**, 323–327.

BLOWER J.G. & GABBUTT P.D. (1964). Studies on the millipedes of a Devon oakwood. *Proc. Zool. Soc. Lond.* **143**, 143–176.

BRZIN M., KOVIC M. & OMAN S. (1964). The magnetic diver balance. *Compt. Rendus Lab. Carlsberg*, **34**, 407–431.

CANCELA DA FONSECA J.-P. (1966). L'outil statistique en biologie du sol. III. Indices d'intérêt écologique. *Rev. Ecol. Biol. Sol.* **3**, 381–407.

CASSAGNAU P. (1961). *Ecologie du sol dans le Pyrénées centrales*. Hermann, Paris.

DEBAUCHE H.R. (1962). The structural analysis of animal communities in the soil. In: *Progress in Soil Zoology*. P.W. Murphy (ed.). Butterworths, London. 10–25.

DELANEY M.J. (1957) Life histories in the Thysanura. *Acta Zool. Cracov*, **2**, 61–90.

DHILLON B.S. & GIBSON N.H.E. (1962). A study of the Acarina and Collembola of agricultural soils. I. Numbers and distributoin in undisturbed grassland. *Pedobiologia*, **1**, 189–209.

EDWARDS C.A.T. (1955). Simple techniques for rearing Collembola, Symphyla and other small soil-inhabiting arthropoda. In: *Soil Zoology.* D.K.McE. Kevan (ed.). Butterworths, London. 412–416.

EDWARDS C.A.T. (1958). The ecology of Symphyla. Part I. Populations. *Ent. exp. and appl.* **1,** 308–319.

EDWARDS C.A.T. (1959). The ecology of Symphyla. Part II. Seasonal soil migrations. *Ent. exp. and appl.* **2,** 257–267.

EDWARDS C.A.T. (1961). The ecology of Symphyla. Part III. Factors controlling soil distributions. *Ent. exp. and appl.* **4,** 239–256.

EDWARDS C.A.T. & DENNIS E.B. (1962). The sampling and extraction of Symphyla from soil. In: *Progress in Soil Zoology.* P.W. Murphy (ed.). Butterworths, London. 300–304.

FRANKLAND J.C. (1966). Succession of fungi or decaying petioles of *Pteridium aquilinum.* *J. Ecol.* **54,** 41–63.

GLICK D. (1961) *Quantitative chemical techniques of histo- and cytochemistry,* 1. Interscience Publ., New York.

GOTO H.E. (1961). Simple techniques for the rearing of Collembola. *Ent. Mon. Mag.* **96,** 138–140.

HALE W.G. (1964). A flotation method for extracting Collembola from organic soils. *J. Anim. Ecol.* **33,** 363–369.

HALE W.G. (1965a). Observations on the breeding biology of Collembola. *Pedobiologia,* **5,** 146–152, 161–177.

HALE W.G. (1965b). Postembryonic development in some species of Collembola. *Pedobiologia,* **5,** 228–243.

HALE W.G. (1966a). A population study of moorland Collembola. *Pedobiologia,* **6,** 65–99.

HALE W.G. (1966b). The Collembola of the Moor House National Nature Reserve, Westmorland: a moorland habitat. *Rev. Ecol. Biol. Sol.* **3,** 97–122.

HEALEY I.N. (1965). Studies on the production biology of soil Collembola. Ph.D. Thesis, University of Wales.

HEALEY I.N. (1967a). The population metabolism of *Onychiurus procampatus* Gisin (Collembola). In: *Progress in Soil Biology.* O. Graff & J.E. Satchell (eds.). Vieweg und Sohn, Brunswick. 127–137.

HEALEY I.N. (1967b). The energy flow through a population of soil Collembola. In: *Secondary Productivity in Terrestrial Ecosystems.* K. Petrusewicz (ed.). Polish Academy of Sciences, Warsaw. Volume II, 695–708.

HUGHES R.D. (1954). The problem of sampling a grassland insect population. Thesis, Imperial College, University of London. 1–152.

HUGHES R.D. (1962). The study of aggregated populations. In: *Progress in Soil Zoology.* P.W. Murphy (ed.). Butterworths, London. 51–55.

KEMPSON D., LLOYD M. & GHELARDI R. (1963). A new extractor for woodland litter. *Pedobiologia,* **3,** 1–21.

MACFADYEN A. (1952). The small arthropods of a *Molinia* fen at Cothill. *J. Anim. Ecol.* **21,** 87–117.

MACFADYEN A. (1955). A comparison of methods for extracting soil arthropods. In: *Soil Zoology.* D.K.McE. Kevan (ed.). Butterworths, London. 315–330.

MACFADYEN A. (1961). Improved funnel-type extractors for soil arthropods. *J. Anim. Ecol.* **30**, 171–184.

MACFADYEN A. (1962). Soil arthropod sampling. In: *Adv. Ecol. Res.* **1**. J.B. Cragg (ed.) Academic Press, London. 1–34.

MACFADYEN A. (1963). The contribution of the microfauna to total soil metabolism. In: *Soil Organisms.* J. Docksen & J. van der Drift (eds.). North Holland Publishing Co., Amsterdam. 3–16.

MACFADYEN A. (1968). Notes on methods for the extraction of small soil arthropods by the high-gradient apparatus. *Pedobiologia,* **8**, 401–406.

MILNE S. (1962). Phenology of a natural population of soil Collembola. *Pedobiologia,* **2**, 41–52.

MURPHY P.W. (1962) (Ed.). *Progress in Soil Zoology.* Butterworths, London.

NEESS J. & DUGDALE R.C. (1959). Computation of production for populations of aquatic midge larvae. *Ecology,* **40**, 683–689.

NIELSON Co. (1961). Respiratory metabolism of some populations of enchytraeid worms and free-living nematodes. *Oïkos,* **12**, 17–35.

OSTDIEK J.L. (1961). Fluctuations in populations of Collembola within leaf litter in the Patuxent Research Refuge, Maryland. *Cathol. Univ. Amer. Biol. Studies,* **62**, 1–44.

PHILLIPSON J. (1963). The use of respiratory data in estimating annual respiratory metabolism, with special reference to *Leiobunum rotundum* (Latr.) (Phalangida). *Oïkos,* **14**, 212–223.

PHILLIPSON J. (1964). A miniature bomb calorimeter for small biological samples. *Oïkos,* **15**, 130–139.

POOLE T.B. (1959). Studies on the food of Collembola in a Douglas Fir plantation. *Proc. Zool. Soc. Lond.* **132**, 71–82.

POOLE T.B. (1961). An ecological study of the Collembola in a coniferous forest soil. *Pedobiologia,* **1**, 113–137.

POOLE T.B. (1962). The effect of some environmental factors on the pattern of distribution of soil Collembola in a coniferous woodland. *Pedobiologia,* **2**, 169–182.

RAW F. (1955). A flotation extraction process for soil microarthropods. In: *Soil Zoology.* D.K.McE. Kevan (ed.). Butterworths, London. 341–346.

RAW F. (1956). The abundance and distribution of Protura in grassland. *J. Anim. Ecol.* **25**, 15–21.

RAW F. (1967). Arthropods (except Acari and Collembola). In: *Soil Biology.* A. Burges & F. Raws (eds.). Academic Press, London. 323–362.

RAYNAL G. (1964). Sur la morphologie et la position systématique d'*Hypogastrura tullbergi.* *Bull. Soc. d'Hist. Nat. Toulouse,* **99**, 3–4, 484–503.

RICHARDS O.W., WALOFF N. & SPRADBERRY J.P. (1960). The measurement of mortality in an insect population in which recruitment and mortality widely overlap. *Oïkos,* **11**, 306–310.

RYKE P.A.J. & LOOTS G.C. (1967). The composition of the microarthropod fauna in South African soils. In: *Progress in Soil Biology.* O. Graff & J.E. Satchell (eds.). Vieweg und Sohn, Brunswick. 538–546.

SAITO S. (1965). Structure and energetics of the population of *Ligidium japonicum* (Isopoda) in a warm temperate forest ecosystem. *Jap. J. Ecol.* **15**, 47–55.

SALT G., HOLLICK F.S.J., RAW F. & BRIAN M.V. (1948). The arthropod population of pasture soil. *J. Anim. Ecol.* **17**, 139–150.

SATCHELL J.E. & NELSON J.M. (1962). A comparison of the Tullgren funnel and extraction methods of extracting Acarina from woodland soil. In: *Progress in Soil Zoology.* P.W. Murphy (ed.). Butterworths, London. 212–216.

SLOBODKIN L.B. (1962). Energy in animal ecology. In: *Adv. Ecol. Res.* **1**. J.B. Cragg (ed.). Academic Press, London. 69–101.

STARLING J.H. (1944). Ecological studies of the Pauropoda of the Duke Forest. *Ecol. Monogr.* **14**, 291–310.

STURM H. (1959). Die nahrung die Protura. *Naturwiss*, **46**, 90–91.

TALBOT P.H.B. (1952). Dispersal of fungus spores by small animals inhabiting wood and bark. *Trans. Brit. Mycol. Soc.* **35**, 123–128.

TAYLOR B.J.R. (1965). The analysis of polymodal frequency distributions. *J. Anim. Ecol.* **34**, 445–452.

TAYLOR L.R. (1961). Aggregation, variance and the mean. *Nature*, **189**, 732–735.

TUXEN S.L. (1949). Uber den lebenszyklus und die postembryonale entwicklung zweier dänischer Proturengattungen. *Kgl. da. Vid. Selsk. Biol. Skr.* **6**, 3, 1–49.

TUXEN S.L. (1964). *The Protura.* Hermann, Paris.

VAN DER DRIFT J. (1951). Analysis of the animal community of a beech forest floor. *Tijdschr. Ent.* **94**, 1–168.

VANNIER G. (1965). Enciente refrigérée par modules thermoélectriques à effet Peltier ($+30°C$ à $-40°C$) permettant l'observation directe de la microfaune. *Rev. Ecol. Biol. Sol.* **2**, 489–506.

VANNIER G. & CANCELA DA FONSECA J.-P. (1966). L'échantillonnage de la microfaune du sol. *La Terre et la Vie,* **1**, 77–104.

WIEGERT R.G. (1965). Energy dynamics of the grasshopper populations in old field and alfalfa field ecosystems. *Oïkos*, **16**, 161–176.

WOOD T.G. (1965). Comparison of a funnel and a flotation method for extracting Acari and Collembola from moorland soils. *Pedobiologia*, **5**, 131–139.

ZEUTHEN E. (1947). Body size and metabolic rate in the animal kingdom. *Compt. Rend. Lab. Carlsberg*, **26**, 17–161.

ZEUTHEN E. (1950). Cartesian diver respirometer. *Biol. Bull.* **98**, 139–143.

ZEUTHEN E. (1964). Microgasometric methods: cartesian divers. *Proceedings 2nd Int. Cong. Histo- and Cytochemistry.* Springer, Berlin. 70–80.

ZINKLER D. (1966). Vergleichende untersuchungen zur atmungsphysiologie von Collembolen (Apterygota) und anderen bodenkleinarthropoden. *Zeit. f. vergleich. Physiol.* **52**, 99–144.

13

Aphids

A.F.G. DIXON

Introduction

Aphids can reach very high densities in some soils but nothing is known of their rôle in production and energy flow in soil ecosystems. In grassland soils Salt *et al.* (1948) extracted 2,000–18,000, and Macfadyen (1953) 100–31,400 aphids per square metre; and Dunn (1959) recorded densities of 100–64,500 per square metre for *Pemphigus bursarius* L., the lettuce root aphid, in market garden soils. In addition, the respiratory rate of aphids is rather high and they produce an excretory material, honeydew, which is very rich in sugars. The scant attention paid to aphids in soil ecology is possibly a reflection of the fact that most work has been done on forest soils where aphids appear to be far less important.

Population estimates

Sampling of aphids in soils is achieved by taking cores of soil and extracting the aphids. Macfadyen (1953) in his comparison of the flotation and funnel methods states that flotation is the more efficient for aphids. He also suggests that repellent chemicals, such as gammexane (benzene-hexachloride), DMP (dimethyl phthalate) and DNOCHP (dinitro-ortho-cyclo-hexyl-phenol), might be used to advantage to replace heat and dryness for driving aphids out of soil cores. In taking soil cores the soil structure is likely to be disturbed resulting in the closure of the potential escape routes for non-burrowing insects like aphids. This could result in greatly underestimating the numbers of aphids present, using the funnel method. Thus a flotation method although more tedious is likely to yield more accurate results. The number of samples that should be taken depends on the density of the aphid under investigation. However, as stressed by Macfadyen (1953) a large number of small samples is preferable to a few large ones. It is possible to save a lot of work and time if an initial investigation defines at what depth in the soil one is likely to find most of the aphids. Salt *et al.* (1948) showed for a grassland soil that 86% of the population are in the top 15 cm of the soil. If this relationship could be accurately determined then a sampling programme based on, e.g. cores taken to a depth of 15 cm could very easily be corrected for with a considerable saving in labour.

In taking the samples every care should be taken to avoid damage to the habitat by trampling by the operator in moving from one sampling site to another, and by the removal of too large a number of soil cores. This can be a serious problem where the habitat to be studied is a relatively small one.

The species composition of the aphids in the samples should be determined wherever possible but this might prove difficult if not impossible in certain areas of the world where little or no systematic work has been done on aphids. However, the error in treating all aphids obtained in the samples as one species may not be very great compared with the errors which arise in estimating their numbers by sampling in the first instance.

Aphids in the soil are likely to be highly aggregated due mainly to their association with a particular species of plant, or roots of a particular size, or ants. Sampling to determine their density is therefore difficult. Where the aphid-plant relationship is a relatively simple one, as in many market garden soils, it is possible to sample at random a number of individual plants and determine the number of aphids living on their roots. This can be repeated at regular intervals through the year so that the seasonal changes in numbers can be determined. Where the relationships between aphid and host are unknown the population estimate will have to be based on spatial random sampling. In both cases, due to the highly aggregated nature of the aphid distribution the result will have to be transformed for accurate comparison between areas, and of the same area at different times of the year. Methods of transforming data obtained in sampling are well reviewed by Southwood (1966). However, as sampling errors are likely to be large then a precise transformation is not necessary and can often lead to problems when comparing means based on different transformations. Statistical advice should be sought at the exploratory sampling stage of the investigation.

The density of aphids per unit area can be converted to biomass per unit area by determining the weight of a representative sample of the aphid(s). This should be done each time density estimates are made as the age distribution of the aphid population, and hence the average weight of an aphid, is likely to change with time.

TABLE 1. The percentage of energy used during development from nymph to adult in growth, honeydew production and respiration in four species of aphid.

Aphid	Percentage energy used			Source
	Growth	Honeydew	Respiration	
Aphis fabae Scop.	19·7	70·6	9·7	Banks (1964, 1965) Dixon*
Drepanosiphum platanoides (Schr.)	4·2	90·5	5·3	Dixon* McNaughton*
Tuberolachnus salignus (Gmelin)	5·5	90·9	3·6	Dixon* Llewellyn* Mittler (1958c)
Megoura viciae Buckt.	66·8	13·4	19·8	Dixon* Ehrhardt (1962, 1963)

* Unpublished.

Methods and apparatus recommended for energy flow determinations

From the literature and unpublished results of my own and my co-workers it is possible to present an account of the fate of energy taken in the food of aphids (Table 1).

Honeydew production, except in *Megoura*, accounts for the greatest proportion of the incoming energy. *Acyrthosiphon pisum* (Harris) also produces little honeydew (Mittler and Sylvester, 1961) and shows the same ratio of 5: 1 for energy expended in growth to energy passed out in the form of honeydew. Although none of these aphids is a root living form it is likely that the same sort of situation holds for them. Where root living aphids are ant attended, which is the usual situation, it is likely that honeydew production will be very high and account for a large proportion of the incoming energy. Banks and Nixon (1958) have shown that when *Aphis fabae* is attended by the ant, *Lasius niger* L., its rate of honeydew production is doubled which would result in the energy used in honeydew production accounting for 83% of that ingested. Where aphids are not ant attended then large quantities of honeydew in the immediate vicinity of the aphid could prove hazardous. It is very sticky and would trap the aphids, or fungus growing on it would smother the aphid colony. In such situations it is noticeable that honeydew production is, in fact, low and the aphid often produces large quantities of a powdery wax which is excellent defence against the sticky honeydew. In such circumstances the amount of energy involved in honeydew production is likely to be low and of the same order as that shown by *Megoura* (Table 1). Some estimate of honeydew and/or wax production is therefore essential along with estimates of production and respiration in determining the rôle of aphids in energy flow in soil ecosystems.

For more sophisticated studies it is possible to obtain a direct measure of food ingested and assimilated, rather than indirect ones from energy used in growth, honeydew production and respiration. The stylets by which the aphid taps the phloem elements of its host plant and obtains its food can be cut and a sample of the food ingested collected directly. This can be achieved by anaesthetizing the aphid in a stream of carbon dioxide and then either cutting the stylets using a guillotine of the type used by Mittler (1958) or a fine pair of scissors as designed and used by Van Soest (1955). On severing the stylets a droplet of fluid will form at the cut ends of the stylets and this can be removed with a micropipette and providing the atmosphere around the cut stylets is kept moist, fluid will continue to flow through the stylets for some time so that reasonable quantities of phloem sap can be collected. This is rather a difficult technique to master and the percentage success with small aphids feeding on non-woody tissues like fine roots is likely to be very low.

Culturing techniques

For the study of excretory, growth and respiratory rates of the aphid(s) concerned it is essential that cultures of the aphid(s) be established in the laboratory. The aphids reared in the laboratory should be compared with specimens collected in the field from time to time during the study so that a correction can be made for any changes that may occur in the aphids as a result of rearing them under laboratory conditions.

The host plant can be grown in soil between two plates of glass (Fig. 1B) so that by removal of one piece of glass aphids can be placed on and removed from roots growing against the glass, or alternatively for larger plants the side of a polythene plant pot can be cut away and replaced by a piece of glass (Fig. 1C, Erhardt and Kloft, 1962). To prevent light disturbing the aphids the glass should be masked to exclude light. Plants for culturing subterranean aphids can also be conveniently grown in culture boxes of the type designed

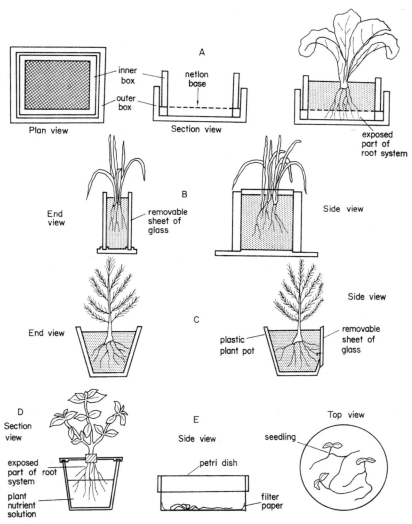

Figure 1. Various methods of growing plants so that aphids can be cultured on their roots. A. Box method (McLean & Kinsey, 1961). B. Plant grown between sheets of glass which can be removed to expose the roots. C. Part of plastic plant pot cut away and replaced by a sheet of glass which can be removed to expose the roots (Ehrhardt and Kloft, 1962). D. Part of the root system immersed in plant nutrient solution. E. Very young seedlings grown on damp filter paper in a Petri dish.

by McLean and Kinsey (1961) for the mass culturing of the lettuce root aphid (Fig. 1A). A rooting medium is placed in both the upper and lower boxes, and the plant is placed in the upper one. When the plant is well established the upper box can be removed and the soil adhering to the roots that have grown through the netlon base should be carefully removed by washing. The lower chamber from which the soil is removed should be kept moist by the addition of damp filter paper to prevent withering of the roots. Some plants can be grown by immersing part of their root system in a nutrient solution (Fig. 1D). The exposed part of the root system can be used for rearing aphids. This method has the advantage of avoiding the use of soil in which aphids are very easily lost. Its one great disadvantage is that many plants cannot tolerate continuous submersion of their roots in water and it is therefore only satis-factory for some plants.

It is also possible to culture root feeding aphids on very young seedlings. The lettuce root aphid can be reared on lettuce seedlings which are only two days old and consist of only cotyledons and a radicle (Fig. 1E) (Judge, 1968). These seedlings can be placed in a Petri dish the base of which is covered with filter paper moistened with a nutrient solution. The seedlings should be kept in the light preferably at the sort of temperature prevailing in the soil. Culturing aphids in this way has the advantage of requiring little space and is easily controlled. However, it does depend upon an abundant source of viable seed.

Production

After obtaining an estimate of the standing crop of aphids the next step in the analysis is to determine the potential production of aphid biomass between one sampling and the next. This is necessary as the aphids reproduce partheno-genetically continuously through the summer months and many of them will be eaten by predators and parasitized before the next sampling period. Just what proportion of the population is removed in this way in subterranean aphids is unknown. An estimate of the reproductive and growth rates of he aphid can be obtained by culturing the aphid in the laboratory on plants grown as illustrated in Figure 1. Rearing the aphid in the laboratory on seedlings and plant roots at temperatures prevailing in the soil will enable one to determine the speed at which the aphid develops and the number of offspring it produces per unit time and its length of life. The clip cage illus-trated in Figure 2 is very useful for isolating aphids on roots for growth and reproduction studies. From the estimate of the potential population growth, and knowing the population realised a correction can be made for mortality suffered by the population due to predators and parasites. However, great care should be exercised in extrapolating the results obtained under laboratory conditions to those prevailing in the field. The errors can be reduced if the age of the plant on which the aphids are cultured is comparable with those in the field and the aphids are reared in groups of the same size as those found under field conditions, as aphids can have marked influences on one another resulting in lowered reproductive rate and the production of alate forms which leave the colony and fly off and colonize other plants. Another way of determining the rate of increase of the population is to use the method devised by Hughes (1962) for *Brevicoryne brassicae* (L.). For this one has to know the length of time spent in each instar and also be able to place an individual collected in the field in its appropriate instar. If the first three instars are of the same duration and one can assume that the instars are all equally susceptible to mortality then it is possible if there is a stable instar distribution to calculate the rate of increase of the population. A combination of these two methods may supply the answer.

The calorific equivalent of aphid tissue is likely to be of the order of 4·6 kcals/gm which is the value for birch aphid (*Euceraphis punctipennis*) tissue.

Honeydew

This can be collected by clip caging aphids individually on roots and catching the honeydew they produce in an aluminium foil cup of known weight placed inside the cage (Fig. 2). At the end of the experiment the aphid's wet weight and the weight of the cup plus dry honeydew are determined. The duration of the experiment can only be decided by trial and error. If the cages are left too long the aphid in wandering around the cage will become trapped in its own honeydew. The relationship between aphid weight and honeydew production is likely to be similar to that shown in Figure 3. The variation is due to moulting during which the production of honeydew is discontinued. If the aphid to be studied is ant attended then it is possible to collect the honeydew by making use of the ant. Ants returning from tending aphids have markedly distended abdomens. By taking a sample each of the ants approaching and those returning from an aphid colony it is possible to determine the average quantity of honeydew collected by each ant. The increase in weight shown by those ants returning from an aphid colony is due to the honeydew they have

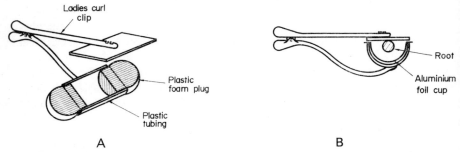

Figure 2. Clip cage for caging aphids individually on plant roots. A. Cage open. B. Cage closed around a root with aluminium foil cup in position for collecting honeydew.

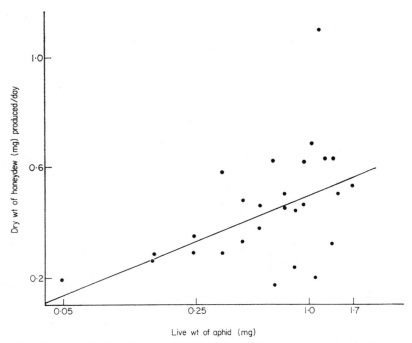

Figure 3. Relationship between body weight and honeydew production in the sycamore aphid, *Drepanosiphum platanoides*.

collected from the aphids. If the number of ants visiting and leaving an aphid colony can be accurately determined then this method has great possibilities for determining the total amount of honeydew produced by aphid colonies.

If a Phillipson micro-bomb calorimeter (Phillipson, 1964) is available then the quantities of honeydew collected by using the aluminium foil cups is sufficient for calorific determinations. However, larger quantities can be collected by placing sheets of aluminium foil below large colonies of aphids and removing the foil several days later. All the aphid bodies and skins are removed from the foil and then the honeydew is washed off with 70% alcohol. The alcohol and water can then be evaporated off leaving the honeydew. This honeydew can be used for calorific determination using more conventional adiabatic or ballistic bomb calorimeter. The calorific equivalent of honeydew is likely to be close to 4·0 kcals/gm., which is the value of sycamore aphid honeydew.

If bomb calorimeters are not available then it is still possible to obtain a good estimate of the calorific value of the honeydew. As the most important constituent of honeydew is sugar it is only necessary to determine the sugar content and then use the calorific value for sugar. The sugar content can be determined with a spectrophotometer, however, in view of the errors already incurred it might be more convenient to use a refractometer. If a refractometer is to be used the sugar content of the honeydew must be determined immediately on collection. For this purpose it can be collected by allowing it to fall on to a waxed surface and then picking up the droplets with a micropipette and transferring them to a refractometer. If the aphid is very small several honeydew droplets will be required. Honeydew collected by ants can be obtained by gently squeezing the ant's abdomen when some of the honeydew will be regurgitated.

In those root dwelling aphids that produce large quantities of flocculent wax as does the lettuce root aphid, *P. bursarius*, some estimate of its production should be made. The wax prevents the aphids from becoming entrapped in their own honeydew or soil water and also affords them protection from parasites and predators. It is also possible that producing wax may also serve as a means of disposing of the excess sugars in their food rather than producing it as honeydew and fouling their immediate environment. This wax is difficult to collect manually because of its flocculent nature but it should be possible to collect it if aphids are reared on small seedlings in Petri dishes and the wax subsequently removed by using a wax solvent.

Respiration

Ehrhardt (1962) and McNaughton (unpublished) have both used the Warburg apparatus for determining the respiratory rate of aphids. However, if the aphid to be studied is a small one, less than 1 mg, many aphids have to be placed in each flask of the Warburg. Phillipson's respirometer (1962) has the advantage of continuous operation over long periods and it can easily be modified to take small animals. However, aphids very quickly suffer from being taken off their host plant so there is no advantage in continuous running over long periods. The apparatus used by myself and co-workers for very small aphids (0·02—0·3 mg) is the Engelmann respirometer (Engelmann, 1963). This has the advantage over the Warburg and Phillipson respirometers in being exceedingly cheap and very easy to construct and operate. In addition, there are no flask constants to determine and the oxygen consumption can be determined directly by reading off the movement of the manometer fluid. It is also very adaptable as the flask in which the aphid is placed can be constructed to a size suitable for the aphid to be experimented with. The smallest flask we use is 0·3 ml. There is no advantage in having the bore of the capillary finer than 0·5 mm as below this the surface tension between the manometer fluid and the walls of the capillary results in a very long lag period before the fluid begins to move and then an erratic movement. All that one needs for the construction and operation of this apparatus is a competent glass blower and a water bath in which the temperature can be kept constant. Interference due to atmospheric pressure changes can be prevented by connecting each respirometer to a large bottle of air maintained at the same temperature as the respirometers. The basic design of the respirometer is illustrated in Figure 4.

The relationship between body weight and respiratory rate is shown in Figure 5. Although none of the aphid species cited in this figure are root living forms the few experimental results available for *Brachycaudus ballotae* (Pass.) which lives on the roots of *Ballota nigra* L., indicate that it is the same as superterranean aphids of the same weight range. The majority of soil dwelling aphids are less than 1 mg in weight and therefore their respiratory rate is likely to be high, and in the region of 1·4–5·8 ml O_2/gm/hr at 20°C.

Discussion

Other arthropod root fluid feeders are cercopids, cicadas, coccids and phylloxerids. They have been reported as reaching very high population

Engelmann constant pressure respirometer

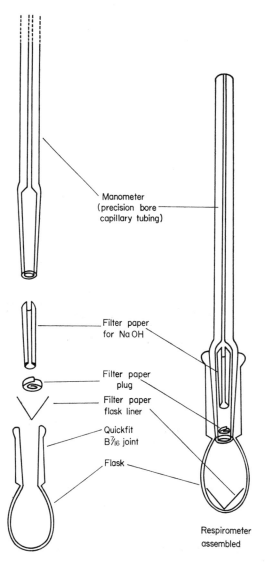

Manometer
(precision bore
capillary tubing)

Filter paper
for NaOH

Filter paper
plug

Filter paper
flask liner

Quickfit
B⁷⁄₁₆ joint

Flask

Respirometer
assembled

Figure 4. Manometer, flask and components of the Engelmann constant pressure respirometer.

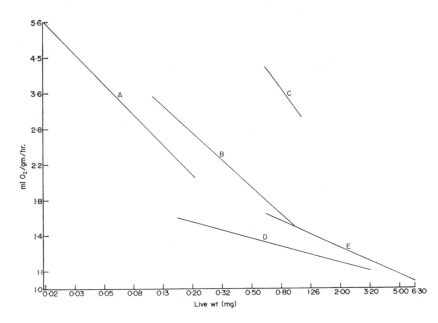

Figure 5. Relationship between body weight and respiratory rate at 20°C for four species of aphid. A. *Eucallipterus tiliae* (Llewellyn, unpublished), B and C. *Drepanosiphum platanoides* nymphs and adults respectively (McNaughton, unpublished), D. *Megoura viciae* (Ehrhardt, 1962), and E. *Tuberolachnus salignus* (Llewellyn, unpublished). [Double logarithmic plot.]

densities and adversely affecting their host plant. Some of them, like the coccids which are ant attended, also produce large quantities of honeydew. Little is known about these groups but it is possible that they could be studied in very much the same way as suggested for aphids.

The main errors lie in determining the density of aphids present in the soil, and the numbers which are being removed from the population by the activity of predators and parasites. Large sampling errors may have to be tolerated because of the time and manpower necessary to reduce them; however, production estimates can be greatly improved by increasing our knowledge of the animals' biology. The present accuracy of calorific honeydew and respiratory determinations are more than adequate for our needs.

Acknowledgements

I am particularly indebted to M. Greene, J. Jackson, M. Llewellyn and F.C. McNaughton for permission to use their unpublished results. My thanks are also due to Mrs S. McKay for drawing the figures.

References

BANKS C.J. & MACAULAY E.D.M. (1964). The feeding, growth and reproduction of *Aphis fabae* Scop. on *Vicia faba* under experimental conditions. *Ann. appl. Biol.* **53**, 229–242.

BANKS C.J. & MACAULAY E.D.M. (1965). The ingestion of nitrogen and solid matter from *Vicia faba* by *Aphis fabae* Scop. *Ann. appl. Biol.* **55**, 207–218.

BANKS C.J. & NIXON H.L. (1958). Effects of the ant, *Lasius niger* L., on the feeding and excretion of the bean aphid, *Aphis fabae* Scop. *J. Exp. Biol.* **35**, 703–711.

DUNN J.A. (1959). The biology of lettuce root aphid. *Ann. appl. Biol.* **47**, 475–491.

EHRHARDT P. (1962). Untersuchungen zur Stoffwechselphysiologie von *Megoura viciae* Buckt., einer Phloemsaugenden Aphide. *Z. vergl. Physiol.* **46**, 169–211.

EHRHARDT P. (1963). Untersuchungen über Bau und Funktion des Verdauungstraktes von *Megoura viciae* Buckt. (Aphidae, Homoptera) unter besonderer Berücksichtigung der Nahrungsaufnahme und der Honigtauabgabe. *Z. Morph. Okol. Tiere*, **52**, 597–677.

EHRHARDT P. & KLOFT W. (1962). Eine Einfache Methode zur Lucht von Lachniden für Untersuchungen im Laboratorium. *Proc. XI Int. Kong. Ent. Wien* 1960, **2**, 544–546.

ENGELMANN M.D. (1963). A constant pressure respirometer for small arthropods. *Ent. News*, **74**, 181–186.

HUGHES R.D. (1962). A method for estimating the effects of mortality on aphid populations. *J. Anim. Ecol.* **31**, 389–396.

JUDGE F.D. (1969). Polymorphism in a subterranean aphid, *Pemphigus bursarius*. I. Factors affecting the development of sexuparae. *Ann. ent. Soc. Am.,* **61**, 819–27.

MACFADYEN A. (1953). Notes on methods for the extraction of small soil arthropods. *J. Anim. Ecol.* **22**, 65–77.

MCLEAN B.L. & KINSEY M.G. (1961). A method for rearing the lettuce root aphid. *J. econ. Ent.* **54**, 1256–1257.

MITTLER T.E. (1958a). The excretion of honey-dew by *Tuberolachnus salignus* (Gmelin) (Homoptera, Aphididae). *Proc. R. ent. Soc. Lond.* (A), **33**, 49–55.

MITTLER T.E. (1958b). Studies on the feeding and nutrition of *Tuberolachnus salignus* (Gmelin) (Homoptera, Aphididae). II. The nitrogen and sugar composition of ingested phloem sap and excreted honeydew. *J. exp. Biol.* **35**, 74–84.

MITTLER T.E. (1958c). Studies on the feeding and nutrition of *Tuberolachnus salignus* (Gmelin) (Homoptera, Aphididae). III. The nitrogen economy. *J. exp. Biol.* **35**, 626–638.

MITTLER T.E. & SYLVESTER E.S. (1961). A comparison of the injury to alfalfa by the aphids, *Therioaphi. maculata* and *Macrosiphum pisi. J. econ. Ent.* **54**, 615–622.

PHILLIPSON J. (1962). Respirometry and the study of energy turnover in natural systems with particular reference to harvest spiders (Phalangiida). *Oïkos*, **13**, 311–322.

PHILLIPSON J. (1964). A miniature bomb calorimeter for small biological samples. *Oïkos,* **15,** 130–139.

SALT G., HOLLICK F.S.J., RAW F. & BRIAN M.W. (1948). The arthropod population of pasture soil. *J. Anim. Ecol.* **17,** 139–150.

SOUTHWOOD T.R.E. (1966). *Ecological Methods.* Methuen, London. 391 pp.

VAN SOEST W. (1955). Een instrument voor het afsnijden van de stylletten bij zuigende bladlnizen. *Tijdschr. PlZiekt.* **61,** 60–61.

14

Ants and Termites

M.V. BRIAN

Introduction

The social nature of these insects makes many of the normal soil sampling methods useless. As they live in defended nests with more or less comprehensively defended areas around them the populations are highly aggregated. They may also be quite mobile and in most cases well able to recruit high densities into baited traps. Some are regularly nomadic. Furthermore all ants and termites are polymorphic with castes that have different distributions and densities. A variable degree of polyethism also exists and complicates some population measurements. Finally the sexual stages are usually highly periodic in their incidence.

Ideally, both the population and the habitat are unaltered and undamaged by the estimation procedure but some destructive sampling is usually inevitable. Even more dangerous is the insidious alteration of the habitat by trampling and the establishment of access windows. The former both kills foragers directly and by maiming a wide variety of prey insects assists them in food capture. The latter necessarily improves the thermoclimate.

Population estimation

If distinct colonies are recognizable population density can be estimated by counting colonies in a sample area and estimating their populations separately. If they are not then a direct sampling procedure must be used. This will be described first.

(1) Soil sampling

This method is only rarely suitable for social insects on account of their high degree of aggregation. However, if clustering is not too marked and enough samples are taken and provided due attention is paid to the statistical principles relevant to asymmetric frequency distributions there is no reason

247

why it should not be tried. The best sample size (a matter primarily of practical convenience) and the number needed can be decided only after a preliminary survey of the population distribution and density.

The coring instrument should be quickly inserted as most social insects live in permanent galleries and will either run away from or collect round a disturbance making the sample non-random. Sands (1965) in sampling termites in soil mounds fixed his corer on the end of a club which he swung swiftly into the base of each mound. The soil core can usually be broken up by hand in a bowl or bucket if the top is painted with polytetrafluoroethylene (fluon), a substance which is so slippy that most insects cannot climb over it. The insects can then be sucked up into a tube or other container and counted. As many ants emit irritating chemicals when disturbed, an electrically driven suction pump is desirable; a powerful commercial vacuum cleaner driven by electricity generated by a small portable petrol-driven machine is quite convenient. If there are many individuals in the sample it may be necessary to narcotize them with carbon dioxide (most usefully stored in cylinders) or to kill them if it is not intended to return them to the nest. Ether and chloroform do not seem to be as harmless as carbon dioxide for narcotization: recovery is slow and often incomplete. If the samples are to be taken away and the insects extracted by flotation or dry heat then the usual precautions for preventing escape or death must be taken. Soil sampling gives figures that can be immediately converted to numbers of individuals per unit area, and, from a knowledge of their average weight, to biomass (g/m^2).

Lasius alienus populations in the soils of heathland in England have been estimated, though not very satisfactorily, by using cores 5·5 cm diameter and 10 cm deep. The mean number of worker ants per core was $1·02\pm0·25$ and 236 samples were taken (Brian, Hibble and Stradling, 1965).

(2) Nest excavation

Small colonies are easily and no doubt accurately estimated in this way, but large and especially diffuse ones are not. It is easy to be persuaded that the whole colony has been discovered when this is far from being the case and this method must not be regarded as a test for the accuracy of other techniques as has sometimes been done. To get the population density and biomass the colony density has of course to be obtained as well.

The simplest time to excavate is when the colony is dormant but, of course, it may not then provide the information required. Most colonies can be located by observation but hidden ones may easily be found by sprinkling food particles and seeing where they are taken. Excavation is then best done gradually either from above or, after digging a trench, from the side (see, for example, Baxter and Hole, 1967). In addition to the usual digging tools sharp-pointed knives of various shapes and shears and secateurs or saws for cutting vegetation are essential. Tracing galleries is easier if a thin suspension of plaster of Paris (anhydrous calcium sulphate) in water is occasionally irrigated or if dry talc (dehydrated magnesium silicate) coloured with a dye, if necessary, is blown in (as by Greaves, 1962). Liquid latex poured or injected into nests sets into rubber and is particularly good for tracing galleries (Brian and Downing, 1958); the insects do, however, stick in it and its use for collecting colonies is limited. The insects can often be sucked up as they emerge but it may be necessary to kill or narcotize them. If so, it is a good idea to do this by infiltrating or injecting the nest sometime beforehand. Hydrocyanic acid (in its commercially available form) or any of the standard soil fumigants might be tried. Those doryline ants which rest overnight above ground can be collected without excavation but it is extremely difficult to get the whole colony and Rettenmeyer (1963) had to guess how much he lost. Those he caught he killed and measured volumetrically.

There are a number of sources of error in this procedure. The individuals that rush out in defence are a special fraction of the community even though they may appear similar. Ant nurses and queens are likely to go away from the encounter carrying brood and may escape completely. There are also, of course, those insects that are away foraging at the time but these will usually return eventually and can be collected. Also it is very difficult not to damage some individuals whilst digging. Finally, many connecting passages are very small and may be blocked with soil and missed; they can go down several metres and are extremely difficult to trace.

Sexuals (and other castes too) can be estimated in some social insects by removing them periodically as they mature through a permanently established approach window such as a plate of glass covered with a slate or other opaque material. This certainly works with many ants but termites will be likely to plaster the glass over; perhaps a chemical coating such as silicone would prevent this. The slate collects solar radiation and if poorly connected to the soil or nest interior it becomes quite warm and attractive in cool climates. Indeed, if used extensively these windows can quite alter the microclimate of the species and have considerable effects on the population. The problem of what to do with the sexuals once removed, counted and weighed if the same colony is being repeatedly censused is a difficult one. They should ideally be put back once all are collected but satisfactory preservation alive is often difficult and there is considerable risk that they will miss the nuptial flights.

It may be hard to decide what the limits of a colony are. Several different brood sites linked by tracks above or below ground are common and a single colony of a mound building species may own several mounds. Part of the population may be in a tree stump or a log, a twig or a fruit on the soil surface and the rest below ground. If there are no obvious physical discontinuities it is essential (and desirable anyway) to test for affinity by mixing samples and watching their reaction but there are cases in ants at least where even non-contiguous and certainly distinct colonies will not fight and it is quite possible that the behaviour will be too indefinite for decisive judgement. If these samples have to be collected from bait it is likely that one colony will have already monopolized it so that a sharp boundary which defines the limit of monopolizing power and not necessarily the limit of foraging, will be obtained. This is true of *Tetramorium caespitum* (Brian, Hibble and Stradling, 1965).

Another method of discovering the boundaries of a colony is to use a radioactive isotope. This may be fed at one point and its limit of spread subsequently plotted either by sampling the insects or, if a strong γ ray emitting isotope is used, by detecting it directly through the soil or nest fabric. Kannowski (1959) used phosphorus 32 on *Lasius minutus* in Michigan U.S.A. and sampled the workers for electron emission; Chauvin, Courtois and Lecomte (1961) used gold 198 to study the food distribution and colony boundaries of *Formica rufa* and *F. polyctena*. Kloft, Holldobler and Haisch (1965) used iodine 131 on *Camponotus* species in Germany and detected the emission directly through wood and soil. Chauvin *et al.* noticed two complications: first, that radioactivity passed from one species to the other in some way and second, that there was a faint natural radioactivity from potassium 40 in the nest material.

Dyes or trace-element chemicals might be tried though their detection necessitates killing a sample of individuals. Trace elements are suggested because methods for detection of small amounts are well established and because they may be absent naturally. Fluorescent dusts that are very useful for flying insects are lost or cleaned off underground.

Many ants and termites forage on or in vegetation above ground as well as in the soil. In calculating the area occupied by a colony of such a species one should take the sum of the vertical projections of the plants used on the soil surface. They may be connected to the nest by trackways whose area is negligible so that an extremely complex shape of foraging range can be created. Combining this figure with the population of the colony gives the economic density or biomass (defined by Elton, 1932).

Owing to the difficulty of deciding exactly what the area is that the colony forages over (and this is an especially acute problem if they are nomadic or make periodic raids to quite distant objectives) it is usually much easier to find the total number of colonies in a given area and calculate Elton's lowest or gross density. Land occupied by other species or not occupied at all will thus be included. Examples of this approach are numerous (see Brian, 1965). Talbot's (1954) work on the ant *Aphaenogaster traetae Forel* on abandoned fields in Michigan U.S.A. is worth special mention.

(3) Traffic assessment

Some colonies of ants are nomadic and move nest from time to time carrying their brood with them. Others may be stimulated to move in various ways, for example, by shading their nest (this has worked with *Myrmica*), or by mechanical and chemical interference. Even so, there are difficulties: one must be sure that a genuine migration is taking place and not simply a raid; some individuals may make several trips between old and new sites; often workers are so densely packed on the trail that they cannot be counted directly and a sample has to be collected. It is also necessary to be in at the start of the migration and stay to the end, assessing flow periodically at various points.

Doryline ants invite the application of this method. Raignier and Van Boven (1955) have made a careful set of estimates of various species of the subgenus *Anomma*. In sampling they inverted a tin over a column and drove it sharply into the ground. It had paper soaked in ether in it and the ants could be collected and counted after immobilization. After averaging many observations on different colonies Raignier and van Boven found that the density of ants in the columns of one species was $7.71/cm^2$, the columns were 2·98 cm wide and they travelled at 169 m/hr for 56·3 hours. This meant that they were 9·5 km long and contained about 22 million ants. Similar methods have been used by Schneirla and Reyes (1966). Rettenmeyer (1963) counted the time taken for 25 ants to pass a line but this became difficult if as many as 7 a second passed and he tried counting in threes.

An interesting way of estimating the active foraging population of Wood ants (*Formica rufa* group) has been tried by Holt (1955). Each nest has a number of tracks to trees on which traffic volume can be measured. The

Chapter 14

theory (which is explained in his paper) depends on the assumption that the traffic is proportional to the population and inversely proportional to the trip time (including both that within the field and in the nest). He counted the flow both ways on all the radials several times and obtained an average. This also established that the flow in and out was about the same and hence that the population at work was stationary. To find the trip time he marked foragers with paint as they left or entered the nest (with practice this could be done without disturbing them) and timed their return. In a case like this loss of the mark only wastes the time of the observer, it does not invalidate any observations that are made. The number of ants that can be marked at a time depends on the number of distinguishable colours and the possibility of marking in different parts of the dorsal abdomen and thorax. His figures were: flow 370 a minute; time out of nest 129 minutes; in nest 31 minutes; and hence the population, 370×160 came to about 60 thousand.

There are as Holt points out many sources of error apart from a not inconsiderable natural variation. These include: a diurnal variation in flow, and foragers not using the tracks.

(4) Petersen's mark-recapture system

The statistical principles involved may be found in Bailey (1951) or in a series of papers by Leslie and Chitty (see Leslie, Chitty and Chitty, 1953) or in Jolly (1963). Southwood (1966) has recently summarized them. This method has been used by various people for estimating ant populations. Others maintain that it is inappropriate as the basic assumption, of random dispersal and random sampling, is unquestionably not upheld in social insects (Holt, 1955; Golley and Gentry, 1964). Nor, in all probability, is it in any animal population though Petersen's method is widely used. It may be possible in some species to split the population into sub-sections, such as foragers, nurses and sexuals, within which movement is random. However, nurses often help forage if bait is available nearby and foragers always rest and unburden in the nest. There is no doubt nevertheless that this method should be more widely tried for ants and termites provided it is used critically and cautiously.

(a) *Sampling.* Preliminary experiments on the dispersal of marked individuals within the society should always be made; if dispersal is imperfect the nest area should be sampled in various places. Permanent access positions can be established with stones or perforated tubes. Baiting with food is not recommended as it both biasses the sample with foragers and decreases their liability of recapture. Deep representative samples should be aimed at, comprising at least a tenth of the population. This method is useful for estimating sexuals provided they can all be done at once; sometimes the earliest have flown before the last emerge and sometimes they leave for nuptials before the recapture sample can be taken and the whole effort is wasted (a mishap that continuous collection avoids).

(b) *Marking*. Ants and termites remove paint; durable marks need to be structural or internal. Cuticular spines may be numerous enough for a few to be expendable without damaging the insect. Scars can also be made though they do not show up well against dark cuticles. Partial or complete amputation of appendages should be avoided as the mobility and activity of the individual is bound to be impaired. Removal of the last tarsal joint of sexual forms is perhaps all right as they can manage inside the nest quite well both before and after mating. Unless handled with care some individuals may lose whole legs and hence their mark and if they are small and numerous cutting off feet can be extremely tedious.

For internal marking certain dyes are useful provided the cuticle is translucent. Neutral red accumulates in the urate cells of the fat body of ant larvae and dyes them permanently red. This can be seen through the skin and is also carried on unchanged into the adult. Radioisotopes overcome opaque cuticles and can be quickly administered though their use does demand some physical knowledge and may be quite expensive both in material and apparatus. Special trials must be devised to show that the isotope is harmless to the insects when used in concentrations high enough to be easily detectable. These may take the form of laboratory cultures of treated and untreated groups that can be compared for longevity, fecundity, growth rate or other useful quantity.

Phosphorus 32 is recommended. It emits electrons vigorously but hardly any of the much more dangerous γ rays and it is consequently easy to handle without major precautions. It can easily be detected outside the insect which is all that is necessary. Moreover it only contaminates the habitat for a short period (with a half life of 14 days it is virtually inactive after 6 months). As an orthophosphate in neutral solution it is chemically acceptable to the organism and permeates its tissues. The individuals to be marked may be dipped in a solution (Odum and Pontin, 1961) or the orthophosphate may be incorporated in syrup or other attractive food (Humus has been used with termites— Alibert, 1963) and then fed to the individuals in small groups so that each feeds directly from the source (Brian, Hibble and Stradling, 1965). This ensures a well-distributed and reasonably controllable intensity of marking. One drawback is the transmission of radioisotope in food, faeces and glandular excretions and exudates the rate of which is customarily measured as the 'biological half-life'. To reduce this effect at least 24 hours must be allowed for ingestion and assimilation before the insects are returned to the nest. During this time, in the case of ants, the crop will empty and regurgitation to unmarked colleagues become impossible but the situation is less predictable in termites (Alibert, 1959). Even though some individuals will, in spite of these precautions, be lightly contaminated they can usually be readily distinguished from the strongly marked originals. The marked insects must be returned to the various places they were drawn from, and adequate period allowed for dispersion (ascertained in preliminary work) and a final recapture sample taken in the same way as before. The radioactive ones in the sample can be found after narcotization with carbon dioxide either by passing them under a Geiger counter (on a moving belt if numbers are large) or by spreading them out on an X-ray plate in an atmosphere of carbon dioxide and producing an autoradiograph. Provided the total number in the recapture sample is known (either by direct count or by estimation after weighing a suitable sample) all the necessary information is available.

In current work on *Tetramorium caespitum* the workers are fed on radioactive syrup (one millicurie in 9 ml syrup) in the laboratory for 3 days, allowed 4 days to drain their crops and returned to the nest for 7 days before sampling again. About 2,000 are marked and about 3,000 recaptured. The populations are in the order of 10,000 and the error about 5% of that.

Population flow

This is, of course, compounded of inflow (fecundity) and outflow from death at all ages. Since the colonies of social insects are virtually closed, migration in and out is negligible and the whole flow of population can be regarded as birth and death (this is not true of the sexuals of course). In ants as distinct from termites the holometabolous life cycle gives a clearly distinguishable adult which enables this class to be considered as the population, and it is often most convenient as a first approximation to do so. In termites, however, this is not so as the adult is difficult to define and the young stages are active and work productively anyway; the best procedure will depend on the characteristics of the species involved.

(1) Inflow

The new adults formed in an ant colony each year can be estimated directly by removing the pupae as they form and replacing them at the end of the season, but this is not very satisfactory as it involves too much interference. Digging the nest up and counting the total brood is not likely to give a very good estimate either as it will be in all stages from egg to adult and eggs in particular are difficult to collect and count without damage. Also, survival from the young larval to the adult stage is known in *Myrmica* in laboratory conditions to be only 2 in 3, Brian (1951). Nevertheless, this has been the only way available and has been used by Golley and Gentry (1964) with the realisation that it is a minimal estimate. Petal (1966) dug up 2 or 3 nests each week and measured the gain in weight of the juvenile population, summing this for the whole season and including the gains of the sexuals produced.

Petersen's mark-recapture method can be applied to the adult population of ants (for details see Ford, 1957; Leslie, Chitty & Chitty, 1953; Jolly, 1963). The marks must last the whole of an individual's lifetime and each age group (normally each year) present in the population at any one time must be distinguishable, that is, differently marked. Anatomical marks like despination can be useful here. In *Myrmica* colonies the workers live at the most two years, so for annual census only 2 marks are needed; either left or right epinotal spine is cut off (Brian, unpublished). Radioisotope marking does not seem to have been used yet for this purpose. One might try cerium 144 with half life 282 days or caesium 134 with half life 2·3 years, both of which were reported to lodge in insects; but all these isotopes need more careful handling than phosphorus 32 as they emit substantial amounts of γ radiation (O'Brien and Wolfe, 1964; Schultz and Klement, 1961; Comar, 1955). Difficulty is likely to arise over providing different marks each year (or age class if other than a year). This might be done quantitatively if the dose is well controlled, year-old individuals being recognized by their comparatively weak signals. Otherwise isotopes with different emission characteristics might be used, for example, an electron emitter one year and a γ ray emitter the next. They could be assessed simultaneously autoradiographically if part of the population is separated from the X-ray plate by an electron impermeable barrier.

The use of this method for the estimation of sexual production has been mentioned already.

(2) Outflow

Petersen's method supplies information on survival too. Laboratory studies have shown (Brian, 1951) that quite a proportion of adult *Myrmica* die shortly after emergence, often from visible defects. Others may survive up to 2 years. This has been confirmed by field studies in which it was found that only a small proportion of workers present one spring are present the next spring as well (unpublished).

The mark-recapture method is applicable to queens but not sexuals as they sooner or later emigrate. Male ants are usually adult for less than a year but this varies. Quantitative data about the survival of sexual female ants is very much needed but very difficult to collect. It is frequently noted that few of the vast numbers produced can establish colonies or even enter existing ones. Marking with radioisotope through the food is easy but owing to wind as well as active flying most leave the zone of origin (there are, of course, contrary cases in which many remain in their natal nest). Indirect methods might be useful at this stage. Thus if the population of queens in an area is known and their longevity ascertained one can calculate how many replacements are necessary if a stationary population is assumed. The total production of sexual females in the same area can be obtained, whence the proportion lost (mainly to predators and accident) can be calculated.

Energy flow

(1) Food intake

Many quantitative estimations of food collection have been made on natural ant populations. The Wood Ant (*Formica rufa* group) of Europe that eats honeydew and insects has been a favourite, and studies have been made on the seed collector *Veromessor* in California by Tevis (1958) or *Formica subuitens* in British Columbia which catches prey on the soil surface and attacks aphids in trees by Ayre (1959) and on the grassland predatory ant *Myrmica* by Petal (1966) in Poland. The leaf cutting ants (Attini) cut and collect a variety of foliage which has been counted and weighed on returning to the nest (Weber, 1966). It is relatively easy though laborious to sample the food intake by collecting returning foragers, removing their burden and replacing them. Sampling should, of course, take into consideration the diurnal and seasonal variables. On the other hand if too much is collected from one colony it will either forage more in an attempt to compensate for the loss or, in the long run, dwindle in size through starvation. Honeydew collection can be estimated by comparing the weights of a sample of incoming with outgoing foragers (as, for example, Holt, 1955). Chauvin (1966) has recently devised a trap applicable to *Formica rufa* type ants in which food is collected from foragers which are made to drop into a box by constructing suitable artificial trackways. They lose their prey during this process and leave the box without it; Leplant (1966) has reported on the first result. This method may be applicable to other ants that have surface tracks provided their nests are not too big and provided that they can be persuaded to drop and lose their prey. Much clearly depends on their behaviour. Although the general feeding habits of termites are now quite well understood (Noirot, 1958; Weesner, 1960; Harris, 1966), quantitative estimates of their food consumption are badly needed.

Collection of the unused fraction of food is much more difficult. Chauvin (1967) has used a circle of gutters round the mounds of *Formica polyctena* into which ants throw their debris. Ants often throw waste indigestible insect and plant remains away in a particular spot (rubbish heap) usually not far from the exit of their nest, and many termites incorporate it in their nest structure (see, for example, Becker and Seifert, 1962). The best way of settling this, however, is to work with small laboratory cultures which can be given known amounts of food and the indigestible part collected. In addition, as the larvae of ants only defecate once, just before they begin metamorphosis, the residues of their entire life can be fairly easily collected. Another feature that should simplify this assessment is the tendency for some species to consume pure fuels (sugars or celluloses as the case may be) as well as preponderately proteinaceous growth foods. Stumper (1961) calculated that 3 calories per gram per day, were consumed by an ant (*Proformica nasuta*) that survives on stored honey during a period of dormancy. Ayre (1960) has measured the food (as fly grubs) needed by *Formica polyctena* under laboratory conditions.

(2) Respiration

Carbon dioxide emission has been measured on small numbers of termites and ants under laboratory conditions (Luscher (1955), Rockstein (1964), Golley and Gentry (1964)). The best method is that perfected by Phillipson (1962). The methods used can be those appropriate for small soil arthropods as long as the following points are taken into account. Most social insects are probably depressed alone and so numbers should be used. A representative sample of the age and caste structure of the society should ideally be taken:

queens, soldiers, workers of several sizes and brood of various sizes. Preferably these should be supplied with non-living food and allowed to settle in the measurement chamber before the experiment is started. If, thereafter, care is taken to avoid disturbance of any kind a reasonable estimate of the basal metabolism of the society may be obtained. Disturbance not only increases respiratory rate enormously but frequently leads to the emission of toxins. Of course much energy is no doubt normally expended in foraging and it seems very difficult to simulate this for even coils of tubing separating food from the nest are hardly near enough natural to be relevant.

To obtain the best results the cultures should be left for several days and this necessitates a system of electrolytic oxygen replacement. Large ants and termites are likely to be more difficult than small ones in these situations and special designs of respirometer chamber will no doubt have to be developed. One should, of course, be careful to avoid having too much soil or organic nest debris unless this is balanced in the compensating chamber, for it respires considerably.

From this, if it is measured over a suitable range of temperature, and if the temperature and the population is known an estimate of the respiratory rate and the heat production of the population can be made. The clear drawback is that quiescent laboratory animals are not the same as foraging, fighting and feeding wild ones. Nor is there one temperature even in homoiothermal colonies nor a series of temperatures that vary spatially and temporally. Perhaps methods now being pioneered for soil respiration can be applied to whole colonies; this should be easiest perhaps with termites that build elaborate and clearly defined nests with special ventilation features.

Temperature measurement technique need not be considered in detail here. Time variations are best integrated. This can be done electronically by electrolysing a solution with a current that is a function of the temperature of a wire or a thermistor inserted in the appropriate place (Macfadyen, 1956). Spatial integration can be achieved by using several probes. Or it can be done chemically by measuring the amount of an organic solvent evaporated and reabsorbed in a hermetically sealed capsule (Brian, 1964). This method, though less precise than the former, is often adequate and has the advantages that large numbers of integrators can easily and cheaply be deployed in different parts of the habitat.

There is no doubt that the temperature inside most nests is higher than ambient and may be very steady (see Raignier, 1948; Noirot, 1958; Greaves 1964). This is partly because of the heat generated metabolically, partly because of special structural arrangements for conservation and also because nests are often placed in sunlit sites. Decomposition of the organic matter or the respiration of cultivated fungi if these exist is probably relatively unimportant.

(3) Assimilation

Methods of estimating the numbers and weights of workers and sexuals have been given in the section on population flow. The energy content of some ants has also been determined by the normal bomb calorimetric methods (Golley and Gentry, 1964; Petal, 1966). The most useful apparatus is that devised by Phillipson (1964). One should remember that the newly formed adult is poor in reserve food materials and that it accumulates them gradually during the first few weeks or months of its adult life. The calorific value certainly varies with the nutritive status of the society. Seasonal variation is mainly reflected in the accumulation of stores before periods of dormancy.

References

ALIBERT J. (1959). Les echanges trophallactiques chez le termite a cou jaune (*Calotermes flavicollis*) étudies à l'aide de radioelements. *C.R. Acad. Sci.* **248**, 1040–1042.

ALIBERT J. (1963). Echanges trophallactiques chez un termite supérieur contamination par le phosphore radio-actif de la population d'un nid de *Cubitermes fungifaber*. *Insectes soc.* **10**, 1–12.

AYRE G.L. (1959). Food habits of *Formica subnitens* Creighton (*Hymenoptera*: *Formicidae*) at Westbank, British Columbia. *Insectes soc.* **6**, 105–114.

AYRE G.L. (1960). Der Einfluss von Insektennahrung auf das Wachstum von Waldameisenvolkern. *Naturwiss*, **47**, 502–503.

BAILEY N.T.J. (1951). On estimating the size of mobile populations from recapture data. *Biometrika*, **38**, 293–461.

BAXTER F.P. & HOLE F.D. (1967). Ant (*Formica cinerea*) pedoturbation in a prairie soil. *Soil Sci. Soc. Amer.* **31**, 425–428.

BECKER G. & SEIFERT K. (1962). Ueber die chemische Zusammensetzung des Nest und Galeriematerials von Termiten. *Insectes soc.* **9**, 273–289.

BRIAN M.V. (1964). Ant distribution in a southern English heath. *J. Anim. Ecol.* **33**, 451–461.

BRIAN M.V. (1965). *Social Insect Populations*. Academic Press.

BRIAN M.V. (1951). Summer population changes of the ant *Myrmica*. *Physiol. comp. et oecol.* **2**, 248–262.

BRIAN M.V. & DOWNING B.M. (1958). The nests of some British ants. *Proc. 10th int. Congr. Entom.* **2**, 539–540.

BRIAN M.V., HIBBLE J. & STRADLING D.J. (1965). Ant pattern and density in a southern English heath. *J. Anim. Ecol.* **34**, 545–555.

CHAUVIN R., COURTOIS G. & LECOMTE J. (1961). Sur la transmission d'isotopes radio-actifs entre deux fourmilières d'espèces différentes (*Formica rufa* et *Formica polyctena*). *Insectes soc.* **8**, 99–107.

CHAUVIN P.R. (1966). Un procédé pour recolter automatiquement les proies que les *Formica polyctena* rapportent au nid. *Insectes soc.* **13**, 59–68.

CHAUVIN R. (1967). Un procédé pour recolter tout ce que les *Formica polyctena* rejettent du nid. *Insectes soc.* **14**, 41–46.

COMAR C.L. (1955). *Radioisotopes in Biology and Agriculture.* McGraw-Hill Book Company.

ELTON C.S. (1932). Territory among wood ants (*Formica rufa* L.) at Picket Hill. *J. Anim. Ecol.* **1**, 69–76.

FORD E.B. (1957). *Butterflies.* Third edition. Collins, London.

GOLLEY F.B. & GENTRY J.B. (1964). Bioenergetics of the southern Harvester Ant, *Pogonomyrmex badius. Ecology*, **45**, 217–225.

GREAVES T. (1962). Studies of foraging galleries and the invasion of living trees by *Coptotermes acinaciformis* and *C. brunneus* (Isoptera). *Aust. J. Zool.* **10**, 630–651.

GREAVES T. (1964). Temperature studies of termite colonies in living trees. *Aust. J. Zool.* **12**, 250–262.

HARRIS W.V. (1966). The role of termites in tropical forestry. *Insectes soc.* **13**, 255–266.

HOLT S.J. (1955). On the foraging activity of the Wood Ant. *J. Anim. Ecol.* **24**, 1–34.

JOLLY G.M. (1963). Estimates of population parameters from multiple recapture data with both death and dilution—a deterministic model. *Biometrika*, **50**, 113–128.

KLOFT W., HOLLDOBLER B. & HAISCH A. (1965). Traceruntersuchungen zur Abgrenzung von Nestarealen Holzzerstorender Rossameisen (*Camponotus herculeanus* L. und *C. ligniperda* Latr.). *Ent. exp. & appl.* **8**, 20–26.

KANNOWSKI P.B. (1959). The use of radioactive phosphorus in the study of colony distribution of the ant *Lasius minutus. Ecology*, **40**, 162–165.

LEPLANT J.P. (1966). Répartition systématique des proies rapportées par les fourmis et recoltées dans l'appareil automatique de Chauvin. *Insectes soc.* **13**, 203–216.

LESLIE P.H., CHITTY D. & CHITTY H. (1953). The estimation of population parameters from data obtained by means of the capture–recapture method. 3. An example of the practical applications of the method. *Biometrika*, **40**, 137–169.

LUSCHER M. (1955). Der Sauerstoffverbrauch bei Termiten und die Ventilation des Nestes bei *Macrotermes natalensis* (Haviland). *Actatropica*, **12**, 289–307.

MACFADYEN A. (1956). The use of a temperature integrator in the study of soil temperature. *Oïkos*, **7**, 56–81.

NOIROT C. (1958). Remarques sur l'écologie des termites. *Annls. Soc. r. Zool. Belg.* **89** 151–169.

O'BRIEN R.D. & WOLFE L.S. (1964). *Radiation, Radioactivity, and Insects.* Academic Press.

ODUM E.P. & PONTIN A.J. (1961). Population density of the underground ant, *Lasius flavus*, as determined by tagging with P32. *Ecology*, **42**, 186–188.

PETAL J. (1966). Productivity and the consumption of food in the *Myrmica laevinodis* Nyl. population. *IBP/PT Symposium on Productivity, Warsaw.*

PHILLIPSON J. (1962). Respirometry and the study of energy turnover in natural systems with particular reference to harvest spiders (*Phalangiida*). *Oïkos*, **13**, 311–322.

PHILLIPSON J. (1964). A miniature bomb calorimeter for small biological samples. *Oïkos*, **15**, 130–139.

RAIGNIER A. (1948). L'économie thermique d'une colonie polycalique de la fourmi des bois, *La Cellule*, **51**, 281–368.

RAIGNIER A. & BOVEN J. (1955). Etude taxonomique, biologique et biométrique de Dorylus du sous-genre Anomma (*Hymenoptera formicidae*). *Annls. Mus. r. Congo belge, Sciences Zoologiques*, **2**, 1–359.

RETTENMEYER C.W. (1963). Behavioural studies of army ants. *Kans. Univ. Sci. Bull.* **44**, 281–465.

ROCKSTEIN M. (1964). Editor. *The Physiology of Insects*, **3**. Academic Press.

SANDS W.A. (1965). Mound population movements and fluctuations in *Trinervitermes ebenerianus* Sjostedt (*Isoptera*, Termitidae, Nasutitermitinae). *Insectes soc.* **12**, 49–58.

SCHNEIRLA T.C. & REYES A.Y. (1966). Raiding and related behaviour in two surface-adapted species of the old world doryline ant *Aenictus*. *Anim. Behav.* **14**, 132–148.

SCHULTZ V. & KLEMENT A.W. (1961). Radioecology. *Proc. 1st Nat. Symp. Radioecology*, Colorado State University.

SOUTHWOOD T.R.E. (1966). *Ecological Methods*. Methuen.

STUMPER R. (1961). Radiobiologische untersuchungen uber den sozialen Nahrungshaushalt der Honigameise *Proformica nasuta* Nyl. *Naturwiss.* **48**, 735–736.

TALBOT M. (1954). Populations of the ant *Aphaenogaster* (*Attomyrma*) *treatae* Forel on abandoned fields on the Edwin S. George Reserve. *Contr. Lab. vertebr. Biol.* **69**, 1–9.

TEVIS L. (1958). Interrelations between the harvester ant *Veromessor pergandei* (Mayr.) and some desert ephemerals. *Ecology*, **39**, 695–704.

WEBER N.A. (1966). Fungus-growing ants and soil nutrition. Monogr. 1, Progressos en biologia del suelo. *Actas del Primer Coloquio Latinoamericano de Biologia del Suelo* (*Bahia Blanca, Argentina*). UNESCO, Montevideo.

WEESNER F.M. (1960). Evolution and biology of the termites. *A. Rev. Ent.* **5**, 153–170.

15

Other Arthropods

J. PHILLIPSON

Introduction

This chapter heading is clearly intended to cover the miscellany of soil dwelling arthropods not yet dealt with in this volume. As the handbook's aim is to outline methods for the quantitative study of population, production and energy flow, one must indicate at the outset the relevant 'other arthropod' groups; these are listed below.

ONYCOPHORA	INSECTA	ARACHNIDA
CRUSTACEA	Orthoptera	Scorpionida
Copepoda	Dermaptera	Phalangiida
Ostracoda	Embioptera	Ricinulei
Isopoda	Psocoptera	Palpigradi
Amphipoda	Hemiptera	Amblypygi
Decapoda	Thysanoptera	Uropygi
MYRIAPODA	Neuroptera	Chelonethi
Chilopoda	Mecoptera	Solifugae
Diplopoda	Trichoptera	Tardigrada
	Lepidoptera	
	Coleoptera	
	Hymenoptera	
	Diptera	

Each of these groups has representative species which at some time in their life history are associated with the soil and are dependent upon it for survival. This is not to say that all of them are ubiquitous in soil, e.g. the moisture requiring Ostracoda, Copepoda, Amphipoda, and Trichoptera are limited to forest soils but range from cold temperate regions to the tropics; the Onycophora are only found in warm temperate and tropical forests; and Scorpionida, Ricinulei, Palpigrada, Pedipalpi and Solifugae are restricted to the warmer regions of the earth. Unfortunately little is known about the quantitative aspects of the ecology of some of the groups, particularly in the tropics, and it is not possible as yet to assess their importance in the economy of the soil. In Europe Bornebusch (1930) and Drift (1951) made assessments of the relative importance of the different groups comprising the fauna of forest soils, whereas Dunger (1968) concentrated on the soil fauna of areas reclaimed after open cast coal mining. In Chile a similar approach was made by Di Castri (1963) who studied several soil types ranging from forest through savannah to arable land.

Soil contains a rich and diverse arthropod fauna but it is generally accepted that the most numerous and widespread of the soil arthropods are the Acari and Collembola (see Chapters 11 and 12). Other arthropod groups recognized as important (Drift, 1951; Raw, 1967) are Isopoda, Chilopoda, Diplopoda, Lepidoptera, Coleoptera, Diptera, Isoptera and Hymenoptera (ants); for an account of the last two groups see Chapter 14. Whether one or more of the above-mentioned six groups of 'other arthropods' is always the most important in soil requires further investigation.

Population

Extraction

Before a sampling programme can be designed to assess the population density of a particular species it is essential to have some means of extracting or collecting individuals from the soil. Clearly no one method of extraction will be equally efficient for all 'other arthropods' and it is perhaps pertinent at this stage to point out that a single extraction method may not even suffice for a single species, e.g. eggs and pupae will probably require a different technique to larvae, which in turn may have to be extracted in a manner different to that needed for adults.

The primary object of any extraction procedure is that animals are efficiently separated from the solid, non-living, materials of the soil, and in such a way that they can be readily sorted, identified and counted. Moreover, as shown in Chapter 10, a single life stage of a single species may be extracted most efficiently from one type of soil by a particular method but when that species occurs in yet other soil types quite different extraction procedures might be more appropriate. The problem of extraction is complicated further when one realises the variety of soils that occurs throughout the world (Lutz and Chandler, 1946; Kubiena, 1953; and Simonson, 1957), some mainly mineral (e.g. latosols), some highly organic (e.g. chernozems), some with large particle size but small crumb structure (e.g. old grassland soils).

In selecting methods for the extraction of 'other arthropods' from soil several considerations arise before a decision can be reached. Is the general aim of the study to
(1) assess the population densities of several species within a single soil type?
(2) compare the population densities of several species in different soil types?
(3) assess the population density of one species within a single soil type?
(4) compare the population densities of one species in different soil types?

If the general aim is to assess the population densities of several species within one or more soil types (1 and 2), and if the available personnel is limited, then time will restrict the number of extraction methods that can be profitably employed. It is known that different species, and different life stages of the same species, are extracted with varying efficiencies by any one method and yet some compromise must obviously be reached. Chapter 10, Table VII, shows that in predominantly organic soils behavioural methods of extraction are best for the active stages (i.e. non-egg, -diapause, -pupal stages) of the ten 'other arthropod' groups mentioned (Isopoda, Chilopoda, Psocoptera, Hemiptera, Thysanoptera, Coleoptera, Diptera, Araneida, Chelonethi (=Pseudoscorpionida)). The recommended extraction methods are, in order of merit, the Macfadyen air-conditioned funnel, the simple funnel with heat, and the Kempson infra-red extractor. With predominantly mineral soils the Macfadyen air-conditioned funnel is recommended for the ten mentioned 'other arthropod' groups, the simple funnel with heat and the Kempson extractor being less effective with Coleoptera and Diptera. Inactive life stages (eggs, pupae, etc.) are best extracted by mechanical methods such as the Salt & Hollick and Edwards & Heath flotation methods, or the Aucamp grease film extractor. These mechanical methods are also suitable for extracting Coleoptera and Diptera from mineral soils.

With regard to the above recommendations it should be stressed that they are based on studies of temperate soil types and that only 10 of the 31 mentioned 'other arthropod' groups were enumerated. In drier regions, where the soil may be finely textured and frequently dusty, it is often impossible to use standard behavioural methods in that a mesh size small enough to stop mineral particles falling into the collecting vessel also restricts the movement of animals through the mesh. It is of interest to note, therefore, that Strickland (1947), Salt (1952, 1955), Belfield (1956) and Block (1970) all used flotation techniques in their studies on tropical soil fauna. It can be concluded that for general extraction of the active stages of common, true soil dwelling, 'other arthropods', those which occur in soils of large crumb structure or high organic content are best extracted by behavioural methods, whereas those occurring in mineral soils of small crumb structure are best extracted by mechanical methods. Mechanical methods should be used for all inactive stages.

Following such generalizations one must not forget the less important (?) groups, for example the hydrophilic Copepoda, Ostracoda and Tardigrada, which, whatever the type of soil in which they live, are probably best extracted by mechanical methods; although a wet behavioural method (e.g. Baermann funnel) might well serve in the case of Copepoda and Tardigrada. Other groups such as the Decapoda, Scorpionida, and adult Coleoptera might be better enumerated by direct counts or mark/release/recapture methods (for methods see Southwood, 1966). Yet others because of their susceptibility to mechanical damage (e.g. Phalangiida) may require hand sorting on site. One can only reiterate that no one method is equally suitable for all groups and that when a general extraction method is adopted then some loss of extraction efficiency is inevitable.

The situation is much simpler when the general aim of the study is either to assess the population density of one species within a single soil type or to compare the population densities of one species in different soil types (3 and 4). Here one can take into account the soil type and the species before choosing a suitable extraction technique. Potential researchers would be well advised to know something of the biology of the organisms which they hope to investigate before deciding to use a particular extraction technique.

It is impossible in such short space to provide details of all thirty-one 'other arthropod' groups with which this chapter deals and the reader's attention is drawn to the general accounts of soil fauna by Kühnelt (1961), Kevan (1962), Dunger (1964), Burges and Raw (1967) and Wallwork (1970). These works provide general background information and will lead readers into the literature, as will the collections of papers to be found in Doeksen and Drift (1963), Graff and Satchell (1967) and Phillipson (1970). Well-known works which include descriptions of the numerous extraction methods available for the study of soil animals are Kevan (1955), Macfadyen (1962), Murphy (1962) and Southwood (1966). It should be borne in mind, however, that the majority of methods were designed for use in temperate regions and that they may not always be suitable for certain soil types of non-temperate regions.

The reader should be aware of the variety of animals found in the soil because the particular area in which he chooses to work may well have some of the lesser known groups as important elements of the fauna. Clearly it is not practicable to cover in detail the known ecology and extraction methodology of all the 'other arthropod' groups but some indication of these aspects might prove useful. The remainder of this section therefore accords with the sequence of 'other arthropod' groups listed in the introduction.

Onycophora—the family Peripatidae is mainly equatorial in distribution (the East Indies, the Himalayas, the Congo, the West Indies and northern South America), whereas members of the family Peripatopsidae are limited to south temperate regions (Australia, New Zealand, South Africa, and the Andes). Most of the 65 known species are confined to humid habitats, such as tropical rain forests or beneath logs, stones and leaves. As adults these carnivores reach sizes ranging from 14 to 150 mm and despite their nocturnalism can presumably be sampled quite readily. Hand sorting or dry sieving of leaf material would be useful for extracting these not very abundant animals.

Copepoda—harpacticoid copopods (approximately 1 mm in length) have been recorded in large numbers from damp soils and woodland litter of tropical, sub-tropical and temperate regions. Dunger (1964) records a species from moss cushions withstanding desiccation for 14 days and hence methods of extraction involving drying out of the sample are probably not to be recommended. Where animals are found in organic soils a wet extraction method such as the Baermann funnel should prove suitable, but in more mineral soils a flotation or elutriation technique might be better.

Ostracoda—it appears that members of this group occur so infrequently in soils (see Lawrence, 1953; Kühnelt, 1961) that they cannot be considered important members of the soil fauna. Presumably some form of flotation or elutriation technique would be the most useful for their extraction, although Bidlingmayer's (1957) sand extractor proved to be very good in this respect.

Isopoda—the members of this group, ranging in size from 2–20 mm, are the most terrestrial of the crustaceans. They are widespread, occurring in both temperate and tropical forests, grasslands and deserts. The group has been fairly well studied and the extraction methods used various, e.g. hand sorting (Saito, 1965), hand sorting and trapping out by means of pit-fall traps (Breymeyer and Brzozowska, 1970), Berlese funnel (Macfadyen, 1961; White, 1968), Tullgren funnel (Paris and Pitelka, 1962), and the Kempson, Lloyd, Ghelardi (K.L.G.) infra-red extractor (Sutton, 1968). The last named author recorded densities of *Trichoniscus pusillus* of over 2,000/m^2 in limestone grassland. Following the recommendations of Chapter 10 it would appear that any one of a variety of methods could be adopted for extracting this group.

Amphipoda—terrestrial amphipods range in size from 10–20 mm and can reach densities as high as 4,000/m^2 in the rain forests of Australia. Saito (1965) collected them by hand sorting but no doubt the extraction methods listed for Isopoda would suffice equally well for Amphipoda.

Decapoda—amphibious crabs up to 15 cm long are widespread in coastal regions, or lands subject to irrigation, e.g. rice paddy fields; whereas true land crabs can move further from water, e.g. Niering (1963) noted eight species as abundant in the forest floor debris of a Micronesian atoll. Almost all of the terrestrial and amphibious crabs dig burrows and hence their population densities could be estimated indirectly by counting burrows; alternatively direct counts by digging out (some burrow to a depth of 2 m) or mark/release/ recapture methods might be used effectively.

Chilopoda—centipedes are widespread and range in size from 10 to 180 mm. They are found primarily in wooded and forested regions throughout the temperate and tropical regions although some species occur in deserts and grasslands. Scutigeromorpha are less closely associated with the soil than other centipedes and occur mainly in the sub-tropics and tropics. The Lithobiomorpha and Scolopendroida are only partially subterranean, whereas the Geophilomorpha are true soil dwellers. Knowledge of centipede populations is limited to temperate regions (Cole, 1946; Dowdy, 1951; Auerbach, 1951; Roberts, 1957; and Wignarajah, 1968). Cole and Auerbach used cryptozoa boards (board traps) whereas Wignarajah using Tullgren funnels recorded maximum densities for a single species in British woodlands of $63/m^2$. From Chapter 10 it can be seen that the recommended methods of extraction are any of the three major behavioural extraction techniques for all soils and also the mechanical methods for the mineral soils.

Diplopoda—the millipedes which range in size from 2 to 150 mm are predominantly wood-land forms and Verhoeff (1934) and Blower (1955) consider the grassland forms to be relict forest fauna. The Polydesmoidae are mainly litter dwellers whereas many of the Iulidae and Glomeridae are active burrowers. According to Lawrence (1953) diplopods are probably as important as earthworms in some African forests, and from temperate regions Dunger (1964) records densities as high as $300/m^2$. Saito (1967) and O'Neill (1968) have sampled them by hand sorting whereas Blower (1970) used Tullgren funnels. According to the findings reported in Chapter 10 the simple funnel with heat is as efficient as any other method for extracting millipedes from any type of soil, although in mineral soils a flotation method might be preferred.

Orthoptera—the common litter dwelling orthopterans are roaches (Blattidae) whereas the fossorial types are the mole crickets (Gryllotalpidae). The roaches are mainly tropical but do occur, sometimes in large numbers, in Europe, Asia, Africa and America. Some orthop-terans are quite large, e.g. *Gryllotalpa* reaches 50 mm in length. It is possible that mark-release/recapture techniques, as used for grasshoppers, might be effective in some cases but in general the methods used for Isopoda, Chilopoda and Diplopoda are recommended.

Dermaptera—these orthopteroid insects belong to eight families and reach lengths of up to 25 mm. Although mainly periodic members of the soil fauna some are truly soil dwellers e.g. *Chelidurella ancanthopygia* in Europe. It is a widespread group covering the whole of the palaearctic and areas as far apart as the Sudan and Sarawak. Depending on the life habits mark/release/recapture and cryptozoa boards, or the methods suggested for Chilo-poda, should prove useful for the assessment of population density.

Embioptera—this is another group of orthopteroid insects, the females usually being apterous. Examples occur in all zoogeographical regions but they are mainly tropicopolitan. They generally live in semi-social groups and inhabit silken tubes under stones and logs and in the upper layers of the soil. Individuals respond to desiccation by wriggling out of their tubes towards moisture, thus, short of direct counts, a dry funnel extraction technique should prove most valuable.

Psocoptera—the members of this group most closely associated with soil are the not very abundant Zoraptera which reach a body length of 3 mm and live a colonial life in decaying wood and humus. They are known from Africa, Ceylon, Java, Bolivia and the southern U.S.A. and are probably best extracted by the methods recommended for European Psocoptera in Chapter 10. The second group of Psocoptera, the Psocidi, reach a length of 5 mm, but only a few species are intimately associated with soil, those that are live gregariously under a canopy of fine silk threads. Suggested methods for their extraction are listed in Table VII, Chapter 10.

Hemiptera—root feeding aphids abound but these and like forms are dealt with in Chapter 13. Apart from aphids the most widely known soil Homoptera are the cicadas which can occur in great numbers in warmer regions, although many species of Cercopoidae also spend their immature stages underground. Few Heteroptera occur in the soil proper but Lygaeidae and Ceratocombidae are frequently found in surface litter. In the southern hemisphere primitive bugs with homopterous affinities (Peloridiidae) are quite common in soil. Cydidnidae are fossorial bugs which occur commonly in Europe and Central America. Murphy (1962a) suggests that the use of a repellent chemical used in conjunction with a dry funnel may still be the best extraction method for soil dwelling Hemiptera.

Thysanoptera—the thrips are small slender-bodied insects of from less than 1 mm up to 8 mm in length. They are generally found in association with decaying plant remains, some being permanent soil dwellers, others merely hibernating or pupating. According to Lewis (1960) adults can be recovered with 80% efficiency by driving them from litter by means of turpentine vapour, and he describes a simple apparatus for this purpose. Other recommended methods are listed in Chapter 10.

Neuroptera—few truly subterranean neuropterous larvae are known, an exception being *Ithone fusca* (Ithonidae) from Australia. Rhaphidioptera, which reach a length of 25 mm and some of which live in leaf litter, occur on all continents bar Australia. Ant lions (larvae of Myrmelionidae) are most abundant in tropical regions but like lace wing larvae (Planipennia) are normally associated with sand and dust. For enumeration purposes it is possible that ant-lion pits could be counted directly but the methods recommended for coleopterous larvae in Chapter 10 would probably be best for Neuroptera in general.

Mecoptera—the carnivorous larvae of this group live in litter and soil, especially around rotting tree stumps; they are not considered abundant but can reach up to 20 mm in length. The Panorpidae and Boreidae are widely spread over the northern hemisphere whereas the Bittacidae occur in most parts of the world except the northern Holarctic. It is suggested that the extraction methods recommended for Coleoptera would work well with this group.

Trichoptera—larvae of a few aberrant species live among forest litter in middle and western Europe, where they reach a length of 7 mm and achieve densities of up to 1,200/m² (Drift and Witkamp, 1959). Hand sorting is probably suitable for these cased forms although as they crawl about in moisture a dry funnel technique may well be best (see Drift, 1963).

Lepidoptera—many adult microlepidoptera hide in leaf litter and large numbers of moths pupate in the soil (Varley, 1970, records densities of *Opheroptera brumata* of 100/m²). Certain noctuid and agrotid larvae hide in the soil during the day but feed on the surface at night. Drift (1951) states that incurvariid larvae of *Adela viridella* are litter feeders, other lepidoptera feed by chewing roots, e.g. *Agrotis segetum* do not come above ground to feed. Dunger (1964) reports larval densities of 5/m² using dry funnel extraction but wet funnel methods may be best for organic soils, and flotation methods for mineral soils.

Coleoptera—a large number of families are commonly associated with soil both as larval and adult stages, e.g. Carabidae, Cerambycidae, Cincindelidae, Cleridae, Curculionidae, Elateridae, Lampyridae, Meloidae, Passalidae, Pselaphidae, Scarabeidae, Staphylinidae and Tenebrionidae. Others such as Coccinellidae may overwinter in litter in aggregations of several thousands. Different families have different importance in the varied regions of the world, e.g. Passalidae and Tenebrionidae are primarily tropical whereas Staphylinidae are mainly north temperate. Individual species range in size from 0·2 to 75 mm and so methods of enumeration need to be varied. For fossorial forms such as the Cicindelidae it may be possible to count burrows; large adult forms may be amenable to mark/release/recapture procedures (Schøtz-Christensen, 1965). For larvae and small adults Chapter 10 recommends dry funnel extraction or flotation but many other methods have been employed, e.g. X-rays for wood borers (Yaghi, 1924; Berryman and Stark, 1962); hot water extraction (Milne, 1958) and flotation for dung dwellers (Laurence, 1954). The choice is wide and will depend upon the particular species, its life stage, and the type of soil in which it occurs.

Hymenoptera—ants are dealt with in Chapter 14 but other hymenopterans are frequently periodic inhabitants of the soil, e.g. Andrenidae build subterranean nests, wasps are strongly represented in sandy soils, many parasitic forms such as the chalcidoids and ichneumonoids are found, and certain larval Cynipoidae cause root galls. Root galls can be counted directly, as can hibernating forms (Alford, 1969). Free living larvae can be extracted by the methods appropriate for Lepidoptera. McLeod (1961), Horsfall (1956), and Read (1958) have designed revolving sieves for extracting pupae.

Diptera—adults frequently occur in the organic horizon of the soil profile e.g. Mycetophilidae, Tipulidae, and Phoridae, and a large number of families are present as larvae (Brauns, 1954). Nematocera are numerous in damp soils, the most important families being the Tipulidae, Bibionidae, Mycetophilidae, Sciaridae, Chironomidae, Ceratopogonidae, and Psychodidae. Brachycera also occur in damp soils, e.g. larvae of Tabanidae, Asilidae, Empididae, Stratiomyidae, Rhagionidae and Therevidae. The Cyclorrapha are represented by larvae Phoridae, Syrphidae, Borboridae, Sepsidae, Scatophagidae, Muscidae, Calliphoridae and Sarcophagidae. Larval size ranges from 1 mm to 35 mm and various methods have been used for their extraction. Following the use of flotation, and a combination of hand sorting and wet sieving, Coulson (1962) reported *Tipula paludosa* egg densities of 18,000/m² and 4,600/m² for early stage larvae. Freeman (1967) also used a wet sieving technique for tipulids whereas Milne *et al.* (1958), working with grassland turves, employed a hot water extraction process. Szabo *et al.* (1967) have estimated mean population densities of *Bibio marci* at 200/m², this, a member of the Bibionidae has been successfully extracted with the sieve/flotation technique of d'Aguilar *et al.* (1957). The sand extractor has also proved efficient for Culicoides larvae (Williams, 1960). It would thus appear that some form of wet sieving or wet funnel method would prove as suitable for the extraction of dipterous larvae as the methods recommended in Chapter 10.

Scorpionida—scorpions are widely distributed in, but largely restricted to, tropical and sub-tropical habitats. They reach between 70 and 100 mm in length and live under rocks, in burrows, or in leaf litter. Certain species exist that possess a toxic venom that can be fatal to man and it is therefore difficult to suggest how one might assess their population density. Some form of remote controlled dry sieve might be useful for sorting leaf litter, whereas burrowing forms might be sampled by the age-old Arab custom of thrusting a stick covered with spittle into any likely burrow or crevice lying within a large quadrat and extracting the individuals which cling to the stick.

Araneida—spiders are ubiquitous arachnids and range in size from 1 mm to 75 mm. Some forms such as the Mygalomorphae, certain Lycosidae and Salticidae form burrows; others spin webs over the soil and litter surface, e.g. Agelenidae, Argiopidae, Pholcidae, Theridiidae, Linyphiidae, Microphantidae and Halmiidae. Yet others such as certain Lycosidae, Thomisiidae, Salticidae and Pisauridae hunt in litter and over the soil surface. In many habitats spiders are numerous. Drift (1951) recorded 232/m² in a beech forest floor, Duffey (1962) found densities as high as 623/m² in *Festuca* grassland, and Cherrett (1964) found densities of 300/m² in a moorland habitat. The last-named author assessed the population density of web spinners by direct counts of the webs, and the non-web spinners were extracted in a lateral extraction apparatus devised by Duffey (1962). Duffey (1962) did not find significant differences in the extraction efficiencies of his lateral extraction apparatus and a Tullgren funnel, but gave good reasons for preferring the former.

Phalangiida—range in size between 3 and 20 mm and are more abundant in woodland litter than elsewhere. Heighton (1964) obtained counts of 167/m² in a British woodland. The same author found that hand sorting in the field was the best method of extraction as the transport of litter samples frequently caused mechanical damage to, and hence disappearance of, the delicate Palpatores. Todd (1949) also used hand sorting and direct observation. The rather more robust Laniatores could no doubt be extracted by the lateral extraction used for araneids, whereas the mite-like Cyphopthalmi could be extracted by the Macfadyen canister extractor or some other dry funnel technique.

Ricinulei—these are the least known of the arachnids. They are 5–10 mm in length and are tropical, sub-tropical litter dwellers which have been recorded from Africa and the American continent from Brazil to Texas. D.H. Murphy (personal communication) has extracted them in reasonable numbers from W. African forest litter by means of dry funnel extractors.

Palpigradi—small arachnids (less than 2 mm long), some of which are tropical (Africa) and others warm temperate (Sicily, Texas and California). They inhabit deep litter and humus where they form burrows. They are extremely delicate forms and flotation or wet funnel extraction are probably the best methods of extraction.

Amblypygi—the members of this tropical and sub-tropical group achieve a body length of from 4 to 45 mm. They hide during the day beneath logs, stones, and leaves. Hand sorting or dry sieving is probably the best method of assessing density, although a mark/release/recapture procedure involving Barber traps might prove useful.

Uropygi—the whip scorpions (2 to 65 mm body length) inhabit situations similar to those occupied by Amblypygi, although a few are desert dwellers. Similar extraction procedures to those outlined for Amblypygi could be used.

Chelonethi—pseudoscorpions reach lengths of up to 8 mm and are found throughout the world from tropical rain-forest to scrub and desert regions, to temperate woodlands and moorlands. They can be locally abundant and Gabbutt (1967) found populations in a beech wood which were consistently over 500/m^2 and reached maximum densities of nearly 900/m^2. The extraction methods suggested in Chapter 10 should be used for this group.

Solifugae—a tropical, sub-tropical group best represented in savannah and desert. They reach lengths of up to 70 mm and Lawrence (1965) states that they prey extensively on termites. Some species live under rocks, others form burrows in the soil. Direct counts of burrows could be attempted, otherwise the methods suggested for Amblypygi and Uropygi should be tried.

Tardigrada—tardigrades are minute (less than 1 mm) and soft-bodied. Some species occur in leaf litter but the majority inhabit moss cushions: no method of quantitative estimation has proved entirely satisfactory but Dunger (1964) reports a high percentage recovery with a Baermann funnel and densities greater than 10^5/m^2. Backlund (1938–39) used centrifugal flotation for their recovery, and they may also be extracted by Oostenbrink's (1954) elutriation technique.

Despite the alleged unimportance of some of the lesser known 'other arthropod' groups it is well to know of their existence for in some, as yet unstudied, habitats it is possible that one or more of them may make a significant contribution to the economy of the soil. What is clear is that 'other arthropods' vary markedly in size, in the habitats they occupy, and in abundance, both within and between groups. It is evident that some groups have not been studied quantitatively at all and it is, therefore, difficult to make specific recommendations regarding extraction methods, yet other groups are well-known and methods for their extraction have been well tried in certain soil types. In the long term, however, practical experience is the best tutor and the extraction methods suggested above can be modified to tailor them to the prevailing conditions in the habitat chosen for study.

Sampling

A sampling programme for the purpose of assessing the density of particular species depends upon:
(1) the activity of the life stage to be sampled;
(2) the type of soil in which the life stage lives;
(3) the size of the life stage to be sampled;
(4) the distribution in space of the life stage to be sampled;
(5) the distribution in time of the life stage to be sampled;
(6) the precision of population density estimate required;
(7) the cost (i.e. time required) of sampling.
Items 1, 2 and 3 are important in choosing the most suitable means of extraction or enumeration, and they have been covered already. Items 3, 4 and 7 affect the size of sample unit required and Grieg-Smith (1957), Healy (1962), Wiegert (1962), Cochran (1963), Southwood (1966), and Berthet in Chapter 11 all deal with statistical aspects of this problem. However, for an adequate investigation into sample unit size, Healy (1962) suggests that a preliminary survey be made in which a fairly large number of sample units of the different possible sizes are taken in a stratified random manner. The mean ($\bar{x}$) and standard deviation (s) can then be calculated for each sample unit size and s can be plotted against $\bar{x}$. The expected value of s for any given value of $\bar{x}$ can thus be determined and if time (cost) is not at a premium then the sample unit size with the narrowest confidence limits can be chosen. Generally a higher level of reproducibility is obtained for the same cost by taking more smaller units than by taking fewer larger ones. However, there is a disadvantage in sampling by small units in that a high proportion of zeros may result at low densities. Such truncation makes analysis difficult and has led to the suggestion that larger sized samples should be taken (Spiller, 1948). Nevertheless, moderate truncations can be overcome by suitable transformations, or better still by the use of non-parametric tests (Siegel, 1956; Fraser, 1957) which, unlike parametric tests, do not assume random distribution of variables.

Despite the recommendation that one should aim at the smallest size of sample unit for the greatest precision one must also bear in mind the cost of each part of the sampling routine. This is usually measured in man hours and will be made up of the time required to take a particular sample and count the animals in it, together with the time spent in moving from one sampling site to the next. During the preliminary survey a record should be kept of the cost of each sample unit so that the optimum sample unit size may be calculated. The optimum sample unit size is that which yields the smallest confidence limits of the mean for a given cost. Wiegert (1962) deals with this aspect in some detail as far as vegetation is concerned but the same principles apply to soil sample units. According to Southwood (1966) the relative net cost for the same precision for each sample unit is proportional to

$$C_u S_u^2$$

where C_u =cost per unit on a common basis, i.e. the cost in time per standard unit of habitat, and S_u^2 is the variance per unit on the common basis. Alternatively the relative net precision of each unit will be proportional to $1/C_u S_u^2$. The higher this value, the greater the precision for the same cost. However, the above author does point out that as population density, and hence variance, is always fluctuating too much stress should not be placed on a precise determination of optimum sample unit size.

Item 6, the precision of population estimate required, will govern the number of sample units required. Debauche (1962), Healy (1962), Macfadyen (1962), Cancela da Fonseca (1965), Southwood (1966) and Chapter 11 of this work all give formulae for calculating the number of sample units needed to achieve a given degree of accuracy and there is little point in developing the theme further. It should be stated, however, that the degree of accuracy to be aimed at in absolute population density studies should be of the order of 10%. Unfortunately, with species that show marked aggregation, are rare, or have a short generation time, one is lucky if one can keep the confidence limits within 30% of the mean.

In addition to affecting sample size Item 4, the distribution in space of the life stage studied, is also important in relation to the pattern of sampling, i.e. the location of sample units. The problem of irregular dispersion patterns occurs with most species but the variance of estimates can be minimized by adopting a stratified sampling programme (Healy, 1962), or a self-weighting one (Wadley, 1952).

Item 5, the distribution in time of the life stage to be sampled, is a further complicating factor, for it may well govern both the time and frequency of sampling. Seasonal timing of sampling is clearly determined by the life cycle of the studied species, but nocturnal/diurnal behaviour may also require that sampling be carried out at different times of the day and night (Southwood, 1966). As for the frequency of sampling it is desirable to try and arrange the sampling intervals in such a way as to correspond with the different phenological phases; only in this way can useful life tables be constructed. Berthet, for instance, points out in Chapter 11 that, with animals having a short generation time, it is preferable to sacrifice precision and take a few sample units at frequent intervals than take many at longer intervals of time.

The problems associated with obtaining reliable absolute population density estimates are great, all one can do is to be aware of the problems and do one's best to meet the stringent statistical requirements. This will not always be possible but in any case figures for the variance associated with the estimates should always be presented.

Biomass

When the ultimate aim of the study is a knowledge of production and energy flow, estimates of population density alone are not sufficient. It is necessary to convert numbers to biomass and this is achieved either with some precision by determining the product of the numbers of each life stage with its mean weight, or more approximately by multiplying the total number of individuals present on one sampling occasion by the mean weight of all individuals on that occasion. In this way the biomass of the population can be calculated. With such a large size range of 'other arthropods' it is not possible to enter into details of how individual weights should be determined, suffice to say that with large animals direct weighing is clearly desirable and both live weights and dry weights should be determined. With smaller animals indirect methods of calculating individual weights may have to be employed; e.g. the use of volume and specific gravity (see Heal, 1970, and Chapter 5), or a generalized equation for a particular group which relates some body dimension to weight (see Chapter 11). However, wherever direct weighing is possible it should be done.

Production

Production has been variously defined but as indicated in IBP News 10 (1968): 'The important point is that, every term, intended to imply some concept about which the precise meaning may be in doubt, should be defined briefly and precisely in each publication in which it is used'.

Ricker (1968) gives two definitions of production, the recommended symbol for which is P.

1. Production (1)—For an individual: increase in biomass during Δt, including the weight of any reproductive products released. For a population: the same, plus the weight of fish [*read animals*] that died during Δt (recruitment absent).
2. Production (2)—Same as production (1), except that the weight of reproductive products released is not included. For an individual fish [*read animal*] production (2) = growth = ΔB.

The first of these definitions accords closely with that recommended in IBP News 10 (1968), viz. 'Biomass of materials digested during a specified time interval (regardless of whether they all survive to the end of that interval), less that respired or rejected.' The second definition from Ricker (1968) is more akin to IBP News 10 (1968) biomass change (ΔB), although for populations the weight of animals that died between sampling intervals is additional.

According to Phillipson (1967) production (Production (1) of Ricker (1968)) can be expressed as:

$$P = P_g + P_r + P_e$$

where P = production, P_g = body growth, P_r = reproductive growth, and P_e = excretion + secretion (including cast exuviae, etc.). This equation is idealistic insofar as it is frequently impractical to separate true excretory and secretory products from egesta.

In detailed studies of production great care must be taken to avoid the inclusion of items such as reproductive growth and exuviae in more than one of the above equation parameters; for example P_g and P_r in the case of eggs and young, and P_g and P_e in the case of exuviae. Certain types of calculation estimate P_g as the difference between initial and final body weights over a whole life span, in this case the weight of young born, or eggs produced, and the weight of cast exuviae should be added to obtain a more accurate estimate of production. Other methods, such as calculating P as the sum of the differences between the initial and final body weights of each life stage automatically include the exuvium as part of the final body weight of each stage. For calculations of the last type the inclusion of P_r as a separate item will depend upon whether reproductive products (e.g. eggs) are voided during a particular life stage, thereby being excluded from the final weight of that stage, and whether the weight of developing reproductive products has been included in the final weights of earlier life stages.

Kaczmarek (1967) has produced a comprehensive article on the possible theoretical approaches to the estimation of secondary production and recommends different approaches for micro-, meso-, and macrofauna. His treatment is too detailed to summarize here and only a few methods will be outlined.

Among small animals (e.g. possibly Tardigrada) it is impractical to study individual biomass changes and use should be made of the reproductive rate and average individual biomass according to the formula:

$$P = N_r \cdot \bar{B}$$

where N_r =the number of individuals born during the time interval Δt and $\bar{B}$ =average individual biomass. Depending on the accuracy required N_r can be calculated using the innate capacity for increase, or, in cruder estimations, by means of:

$$N_r = \frac{n_b}{\tau_m} \cdot \frac{\bar{n}_m}{2} \cdot \Delta t$$

where n_b =number of eggs or newborn per female reproductive period, $\bar{n}_m$ =average number adults observed during time interval Δt (divided by 2 when sex ratio is 1:1, otherwise by the appropriate figure), and τ_m =duration of life of mature adult.

Loss due to mortality can be estimated by the equation:

$$M = N_m \cdot \bar{B} = (Nr \cdot \Delta_n) \bar{B}$$

where N_m =the number of individuals dying, $\bar{B}$ =average individual biomass, and Δ_n =change in population density during Δt.

Another possibility for calculating production by small animals is the formula given by Heal (1970) for Protozoa:

$$P = \frac{N_0 + N_1}{2} \times \frac{1}{D} \times t$$

where P =production, N_0 and N_1 are the initial and final numbers (or biomass) in time t, and D is the period of 'doubling' of an individual.

With larger animals (e.g. many 'other arthropods') assessment of biomass change of individuals can be monitored and hence detailed studies of population production attempted. Saito (1965, 1970) calculated P_g for declining isopod and millipede cohorts by using the formula:

$$P_g = (Nt \times \frac{\Delta B}{\Delta t}) + (\frac{N}{\Delta t} \times \frac{1}{2} \times \frac{\Delta B}{\Delta t}) \dots \dots \dots (1)$$

where Nt =the number of individuals surviving at time t_1, Δt =the time period $t_1 - t_0$, $\dfrac{\Delta B}{\Delta t}$ = the mean weight increment in $t_1 - t_0$, $\dfrac{\Delta N}{\Delta t}$ = the change in cohort numbers in $t_1 - t_0$, and $\frac{1}{2}$ is based on a constant mortality rate deduced from a Deevey (1947) Type II survivorship curve.

Formula 1 reduces to:

$$P_g = (Nt + \frac{\Delta N}{2 \Delta t}) \cdot (\frac{\Delta B}{\Delta t}) \quad \dots \dots \dots \dots \dots (2)$$

When there is an overlap of generations (detected from size frequency distribution data) then the above formula must be applied separately to each generation ($\equiv$cohort). This is not always possible, for in a population with continuous recruitment due to natality the separate generations cannot be distinguished, and yet, in addition to the production by members of the original cohort the production of recruited animals must be taken into account. Thus, the final population density Nt may be greater than N_0 during recruitment periods and hence ΔN becomes a negative quantity. Under these circumstances Bolton (1969) suggests the use of:

$$P_g = [(1 - M) . N_0 \times \frac{\Delta B}{\Delta t}] + [\frac{1}{2} \times M . N_0 \times \frac{\Delta B}{\Delta t}] + [\frac{1}{2} \times (M . N_0 + \frac{\Delta N^1}{\Delta t}) \times \frac{\Delta B}{\Delta t}]$$

where mortality and recruitment are assumed constant over Δt; $M =$ mortality rate over Δt, N_0 is the initial number of animals at t_0, ΔN^1 the change in population numbers can be positive or negative according to the balance between recruitment and mortality. If ΔB is assumed constant for survivors, animals that died, and recruits over Δt then the above formula reduces to:

$$P_g = [(1 - M) . N_0 + (\frac{1}{2} \times M . N_0) + (\frac{1}{2} \times M . N_0) + \frac{\Delta NI}{2 \Delta t}] \times \frac{\Delta B}{\Delta t}$$

$$= (N_0 + \frac{\Delta N^1}{2 \Delta t}) \cdot (\frac{\Delta B}{\Delta t}) \quad \dots \dots \dots \dots \dots (3)$$

Thus equation 3 for a population incorporating recruits is similar to equation 2 for a declining cohort, with the substitution of N_0 for Nt and ΔN^1 which is merely a new interpretation of ΔN in that it can be positive or negative. However, P_g for a population undergoing recruitment can be calculated even more simply by calculating the mean population density ($\bar{N}$)

over Δt and multiplying this figure by $\dfrac{\Delta B}{\Delta t}$, since:

$$P_g = \bar{N} \times \frac{\Delta B}{\Delta t} = \frac{(N_0 + Nt)}{2} \times \frac{\Delta B}{\Delta t}$$

$$= \frac{1}{2} \times (2N_0 + \frac{\Delta N^1}{\Delta t}) \times \frac{\Delta B}{\Delta t}$$

$$= (N_0 + \frac{\Delta N^1}{2\Delta t}) \cdot (\frac{\Delta B}{\Delta t}); \text{ i.e. equation 3.}$$

This method is equally suitable for a declining cohort situation; in this case

$$P_g = N \times \frac{\Delta B}{\Delta t} = \frac{1}{2}(2Nt + \frac{\Delta N}{\Delta t}) \times \frac{\Delta B}{\Delta t}$$

and since ΔN is positive

$$P_g = (Nt + \frac{\Delta N}{2\Delta t}) \cdot (\frac{\Delta B}{\Delta t}); \text{ i.e. equation 2.}$$

The use of $\bar{N}$ and $\dfrac{\Delta B}{\Delta t}$ in calculations of P_g will, of course, be less accurate than application of the earlier Saito (1965, 1970) equations, simply by virtue of the fact that the rate of biomass increment by individual recruits is assumed to be the same as that of adults, this is not so in those calculations where discrete generations are followed.

Other methods of calculating P_g are available (Kaczmarek, 1967; Phillipson, 1967) but the methods outlined are believed to be widely applicable and reasonably precise.

It was mentioned earlier that the inclusion, or not, of P_r and P_e in P_g depends on the manner in which $\dfrac{\Delta B}{\Delta t}$ is calculated. Where P_r and P_e are known not to be incorporated in P_g then reproductive growth (P_r) of a population can be estimated by studying the number of pregnant or gravid females in the population, the mean egg number per female, and the mean weight of an egg. Alternatively, the relationship between egg number and body weight or size can be used (Paris and Pitelka, 1962; Phillipson and Watson,

1965; Saito, 1965). Similarly P_e, as far as exuviae are concerned, can be estimated by using a body size-exuvium weight relationship (Wiegert, 1964).

Where figures are not available for the direct estimation of production, but annual respiration of the population is known, one can use the 'Engelmann Line' (Engelmann, 1966), now modified by McNeill and Lawton (1970), to estimate production indirectly.

Energy flux

Energy flux or, as it is more frequently termed, energy flow, is usually expressed in a generalized form by the equation:

$$C = P + R + F + U$$

where in energy terms:

C. *Consumption* = total intake of food energy during a specified time interval.

P. *Production* = energy content of the biomass of materials digested during a specified time interval (regardless of whether they all survive to the end of that interval) less that respired or rejected.

R. *Respiration* = the energy equivalent of that part of assimilation (the sum of production and respiration) which is converted to heat and lost in life processes.

F. *Egesta* = the energy content of that part of consumption which is not digested.

U. *Excreta* = the energy content of that part of the digested material which is passed from the body. (This does not include reproductive and secretory products).

The methods used for determining the parameters C, P, R, F and U are numerous; it is impossible to outline them all in this contribution but a separate IBP Handbook: 'Methods for Ecological Bioenergetics' (Grodzinski and Klekowski, in press) details many of the methods that can be used. Further, IBP Handbook No. 13 (Petrusewicz and Macfadyen, 1970) explores the principles and methods of studying the productivity of terrestrial animals.

The quantification of C, P, R, F and U under natural conditions is not easy, although as indicated previously, P can be calculated from data obtained during field censuses. The difficulty of measuring energy budget parameters in the field means that one must frequently acquire information by conducting experiments in the laboratory or under semi-natural field conditions. Under these circumstances it is well to remember Macfadyen's (1967) statement:

'The final unknown in so many of these studies is, of course, the extrapolation of laboratory data to field conditions which in many cases may lead to the largest errors of all and make certain refinements irrelevant'. Space is not available to discuss this point in full, suffice to say that field methods for measuring the energy budget parameters of soil dwelling animals are either not yet available or can only be used with confidence with certain species. For example, some authors strongly recommend the use of radionuclides in studies of consumption and assimilation; if this is done then it must be assumed that the absorption of radionuclides is not selective and that assimilation is independent of ingestion rate, Hubbell *et al.* (1965) found these assumptions to be invalid in certain cases. It would appear, despite its drawbacks, that laboratory experimentation is often necessary, but that every possible precaution should be taken to simulate natural conditions in the experimental set-up.

Many factors influence the estimation of the energy budget parameters, e.g. the quality and quantity of food materials, the chemico-physical nature of the habitat, and the age, size, physiological state and activity of the animal itself. Providing the seasonal food habits of the studied species are known then the quality and quantity of food included in the experimental set-up can be arranged to simulate natural conditions. Similarly, if records of the chemico-physical nature of the habitat have been taken, particularly humidity (e.g. by cobalt thiocyanate papers, Solomon, 1957) and above all temperature (e.g. by thermistors, or the Pallman-Berthet technique, Berthet, 1960) then some attempt to simulate field conditions can be made. Temperature in particular is known to affect animal metabolism and researchers are advised to allow a period of acclimation to the experimental temperature (in some cases 7–14 days) before actual measurements are taken. Clearly, body dimensions and/or weight are not too difficult to record, and a watch should be kept for any effects due to physiological state or activity (Phillipson, 1962, 1963).

As stated earlier the methods for determining each of the parameters C, P, R, F and U are too numerous to detail here but some indication of the possible approaches might prove useful.

Consumption (C)—under certain circumstances direct quantitative observation might be possible, as was the case with *Myrmica laevinodis* (Petal, 1967); for example, direct observation might be profitably employed in studying dung consumption by larval Scarabeidae. However, most methods of measuring consumption in quantitative terms requires a knowledge of food type and quality before meaningful experiments can be made. Under such circumstances analyses of gut contents and/or faeces should be used with care, particularly where soft-bodied prey or fluids are known to be taken. In this case precipitin tests (Reynoldson and Young, 1963) should prove useful. Alternatively, food preference experiments (Phillipson, 1960a, 1967) might be used, or radionuclides to trace food webs (Reichle and Crossley, 1965; Coleman, 1970). Given information regarding food type and quality absolute consumption rates can be calculated from a knowledge of percentage assimilation of the food under laboratory conditions and faecal output under natural or near natural conditions (Phillipson, 1960a and b, 1967). A less desirable method is the summation of P and $(F+U)$, as determined in the field, and R as measured under simulated field conditions (Smalley, 1960). Radionuclides have also been used in the quantitative determination of consumption (e.g. Reichle and Crossley, 1967). Where consumption cannot be measured the body size–food intake relationship described by Reichle (1968) could be employed.

Production (P)—methods for the calculation of production are given in an earlier section and will not be repeated here.

Respiration (R)—providing the appropriate precautions mentioned in the earlier part of this section, e.g. diel rhythms, are taken into account, any of the many kinds of apparatus available for the measurement of respiratory rate can be used. Phillipson (1967) lists a number of respirometers, many of which are sold commercially; other non-commercially produced respirometers, e.g. the Cartesian diver, Drastich constant pressure respirometer, and Phillipson electrolytic respirometer) are described in Grodzinski and Klekowski (in press). Where measurement of respiration cannot be undertaken the body size–oxygen consumption relationship outlined by Reichle (1968) could be used as a first approximation. Alternatively, where annual production is known the 'Engelmann Line' (McNeill and Lawton, 1970) could be applied.

Egesta + Excreta $(F+U)$—in most arthropods it is impossible to separate egesta and excreta and so these two parameters are usually considered together and equated with faeces. Estimates of faeces production may be affected by periodicity in defaecation but this can be counteracted by adopting the procedure of Phillipson (1960a). Care should be taken to measure faeces production either from a knowledge of gut volume and turnover time under near natural conditions, or from animals which have fed in the field before being placed in containers under near natural conditions, otherwise serious errors can arise in the faeces production estimates, as was shown by Phillipson (1967) working with *Oniscus asellus*.

It is not sufficient to make measurements of any of the above parameters simply at one time of the day or year and every effort should be made to allow for day/night activity (Phillipson, 1962) and seasonal variations in metabolism (Phillipson, 1963).

Thus, in most cases, production (P) can be calculated from field data and quite frequently faecal production (F+U in 'other arthropods') can be measured under natural conditions or circumstances approaching them. Consumption (C) if not measurable directly can be calculated indirectly from a combination of laboratory and field data, and respiration (R) is nearly always measured in the laboratory. Providing each parameter is determined independently, each under slightly different conditions as described above, then the best check of the validity of the estimates is the balancing of the energy budget equation (see Phillipson, 1967). Clearly the estimation of a parameter by difference is not to be recommended as it is bound to make the equation balance, and this step should only be taken as a last resort.

The solving of an energy budget equation requires all parameters to be expressed in the same units and, as is well known, the energy content of biological materials can be obtained by the use of an oxygen bomb calorimeter or a wet combustion procedure (Ivlev, 1934). Oxygen uptake, as measured directly or calculated from carbon dioxide output and R.Q. values, can be converted to energy units by means of the appropriate oxycalorific equivalents. In ecological energetics literature it has been a common practice to use gram calories or kilocalories as basic energy units; however, because there are many different kinds of calorie, and for the sake of uniformity, it is recommended that the Standard International Unit of energy, the Joule (J), be adopted in expressing all future results. For conversion purposes 1 calorie (thermochemical) $\equiv$ 4·184 J.

Hopefully, the reader of this chapter should now be in a position (when faced with an 'other arthropod' population), to plan and execute a quantitative study of at least one aspect of soil ecology—population, production, and energy flow.

References

D'AGUILAR J., BENARD R. & BESSARD A. (1957). Une méthode de lavage pour l'extraction des arthropodes terricoles. *Ann. Epiphyt.* **8,** 91–99.

ALFORD D.V. (1969). A study of the hibernation of bumblebees (Hymenoptera: Bombidae) in southern England. *J. Anim. Ecol.* **38,** 149–170.

AUERBACH S.I. (1951). The centipedes of the Chicago area with special reference to their ecology. *Ecol. Monogr.* **21**, 97–124.

BACKLUND H.O. (1938–39). Eine methode zur quantitativen Untersuchung der Mikrofauna in Moos, Förna, Trift, und dergl., sowie einige vorläufige Ergebnisse mit dieser Methode. *Memor. Soc. Fauna Flor. fenn.* **14**, 27–32.

BELFIELD W. (1956). The Arthropoda of the soil in a West African pasture. *J. Anim. Ecol.* **25**, 275–87.

BERRYMAN A.A. & STARK R.W. (1962). Radiography in forest entomology. *Ann. ent. Soc. Amer.* **55**, 456–466.

BERTHET P. (1960). La mesure écologique de la température par détermination de la vitesse d'inversion du saccharose. *Vegetatio*, **9**, 197–207.

BIDLINGMAYER W.L. (1957). Studies on *Culicoides furens* (Poey) at Vero Beach. *Mosq. News*, **17**, 292–294.

BLOCK W. (1970). Micro-arthropods in some Uganda soils. In: *Methods of Study in Soil Ecology.* J. Phillipson (ed.). Unesco, Paris. 195–202.

BLOWER J.G. (1955). Millipedes and centipedes as soil animals. In: *Soil Zoology.* D.K.McE. Kevan (ed.). Butterworths, London. 138–151.

BLOWER J.G. (1970). The millipedes of a Cheshire wood. *J. Zool., Lond.* **160**, 455–496.

BOLTON P.J. (1969). Studies in the general ecology, physiology and bioenergetics of woodland Lumbricidae. Ph.D. Thesis, Univ. of Durham, U.K.

BORNEBUSCH C.H. (1930). The fauna of forest soil. *Forst. Fors-Vaes. Danm.* **11**, 1–224.

BRAUNS A. (1954). Untersuchungen zur angewandten Bodenbiologie. Bd 1, Terricole Dipterenlarven, Musterschmidt, Göttingen. 179 pp.

BREYMEYER A & BRZOZOWSKA D. (1970). Density, activity and consumption of Isopoda on a Stellaria-Deschampsietum meadow. In: *Methods of Study in Soil Ecology.* J. Phillipson (ed.). Unesco, Paris. 225–230.

BURGES A. & RAW F. (eds.) (1967). *Soil Biology.* Academic Press, London and New York. 532 pp.

CANCELA DA FONSECA J.P. (1965). L'Outil Statistique en Biologie du Sol. I. Distribution et fréquences et tests de signification. *Rev. Ecol. Biol. Sol.* **2**, 299–332.

CHERRETT J.M. (1964). The distribution of spiders on the Moor House National Nature Reserve, Westmorland. *J. Anim. Ecol.* **33**, 27–48.

COCHRAN W.G. (1963). *Sampling Techniques.* 2nd ed. New York. 413 pp.

COLE L.C. (1946). A study of the cryptozoa of an Illinois woodland. *Ecol. Monogr.* **16**, 49–86.

COLEMAN D.C. (1970). Food webs of small arthropods in a broomsedge field studied with radio-isotope-labelled fungi. In: *Methods of Study in Soil Ecology.* J. Phillipson (ed.). Unesco, Paris. 203–207.

COULSON J.C. (1962). The biology of *Tipula subnodicornis* Zetterstedt, with comparative observations on *Tipula paludosa* Meigen. *J. Anim. Ecol.* **31**, 1–21.

DEBAUCHE H.R. (1962). The structural analysis of animal communities of the soil. In: *Progress in Soil Zoology.* P.W. Murphy (ed.). Butterworths, London. 10–25.

DEEVEY E.S. (1947). Life tables for natural populations of animals. *Quart. rev. Biol.* **22** 282–314.

Chapter 15

Dɪ Cᴀsᴛʀɪ F. (1963). Etat de nos connaissances sur les biocoenoses edaphiques du Chili. In: *Soil Organisms*. J. Doeksen & J. van der Drift (eds.). North Holland Publishing Co., Amsterdam. 375–385.

Dᴏᴇᴋsᴇɴ J. & Dʀɪғᴛ J. ᴠᴀɴ ᴅᴇʀ (eds.) (1963). *Soil Organisms*. North Holland Publishing Co., Amsterdam. 453 pp.

Dᴏᴡᴅʏ W.W. (1951). Further ecological studies on stratification of the Arthropoda. *Ecology*, **32**, 37–52.

Dʀɪғᴛ J. ᴠᴀɴ ᴅᴇʀ (1951). Analysis of the animal community in a beech forest floor. *Tijdschr. Ent.* **94**, 1–168.

Dʀɪғᴛ J. ᴠᴀɴ ᴅᴇʀ (1963). The disappearance of litter in mull and mor in connection with weather conditions and the activity of the macrofauna. In: *Soil Organisms*. J. Doeksen & J. van der Drift (eds.). North Holland Publishing Co., Amsterdam. 125–133.

Dʀɪғᴛ J. ᴠᴀɴ ᴅᴇʀ & Wɪᴛᴋᴀᴍᴘ M. (1959). The significance of the breakdown of oak litter by *Enoicyla pusilla* Burm. *Arch. Néerl. de Zool.* **13**, 486–492.

Dᴜғғᴇʏ E. (1962). A population study of spiders in limestone grassland. *J. Anim. Ecol.* **31**, 571–599.

Dᴜɴɢᴇʀ W. (1964). *Tiere im Boden*. Ziemsen Verlag, Wittenberg Lutherstadt. 265 pp.

Dᴜɴɢᴇʀ W. (1968). Die Entwicklung der Bodenfauna auf rekultivierten Kippen und Halden des Braunkohlentagebaues. *Abh. Ber. Naturkundemus. Görlitz.* **43**, 1–256.

Eɴɢᴇʟᴍᴀɴɴ M.D. (1966). Energetics, terrestrial field studies, and animal productivity. *Adv. Ecol. Res.* **3**, 73–115.

Fʀᴇᴇᴍᴀɴ B.E. (1967). Studies on the ecology of larval Tipulinae (Diptera, Tipulidae). *J. Anim. Ecol.* **36**, 123–146.

Fʀᴀsᴇʀ D.A.S. (1957). *Non-parametric Methods in Statistics*. Wiley, New York.

Gᴀʙʙᴜᴛᴛ P.D. (1967). Quantitative sampling of the pseudoscorpion *Cthonius ischnocheles* from beech litter. *J. Zool., Lond.* **151**, 469–478.

Gʀᴀғғ O. & Sᴀᴛᴄʜᴇʟʟ J.E. (eds.) (1967). *Progress in Soil Biology*. North Holland Publishing Co. Amsterdam. 656 pp.

Gʀɪᴇɢ-Sᴍɪᴛʜ P. (1957). *Quantitative Plant Ecology*. Butterworths, London. 198 pp.

Gʀᴏᴅᴢɪɴsᴋɪ W. & Kʟᴇᴋᴏᴡsᴋɪ R.Z. (eds.) (in press). *Methods for Ecological Bioenergetics*. Blackwell Scientific Publications, Oxford.

Hᴇᴀʟ O.W. (1970). Methods of study of soil protozoa. In: *Methods of Study in Soil Ecology*. J. Phillipson (ed.). Unesco, Paris. 119–126.

Hᴇᴀʟʏ M.J.R. (1962). Some basic statistical techniques in soil zoology. In: *Progress in Soil Zoology*. P.W. Murphy (ed.). Butterworths, London. 3–9.

Hᴇɪɢʜᴛᴏɴ B.N. (1964). Biological studies on certain species of British Phalangiida. Ph.D. Thesis, Univ. of Durham, U.K.

Hᴏʀsғᴀʟʟ W.R. (1956). A method for making a survey of floodwater mosquitoes. *Mosq. News*, **16**, 66–71.

Hᴜʙʙᴇʟʟ H., Pᴀʀɪs O.H. & Sɪᴋᴏʀᴀ A. (1965). Radiotracer gravimetric and calorimetric studies of ingestion and assimilation rates of an isopod. *Health Physics*, **11**, 1485–1501.

IBP Nᴇᴡs 10 (1968). Terminology. *IBP News*, **10**, 6–8.

Iᴠʟᴇᴠ C.S. (1934). Eine Mikromethode zur Bestimmung des Kaloriengehaltes von Nährstoffen. *Biochem. Z.* **275**, 49–55.

KACZMAREK W. (1967). Methods of production estimation in various types of animal population. In: *Secondary Productivity of Terrestrial Ecosystems*. K. Petrusewicz (ed.). Warsaw. 413–446.

KEVAN D.K.McE. (ed.) (1955). *Soil Zoology*. Butterworths, London. 512 pp.

KEVAN D.K.McE. (1962). *Soil Animals*. Witherby, London. 244 pp.

KUBIENA W.L. (1953). *The Soils of Europe*. London. 318 pp.

KÜHNELT W. (1961). *Soil Biology* (English ed., transl. N. Walker). Faber, London. 397 pp.

LAURENCE B.R. (1954). The larval inhabitants of cowpats. *J. Anim. Ecol.* 23, 234–260.

LAWRENCE R.F. (1953). *The Biology of the Cryptic Fauna of Forests*. Balkema, Cape Town, Amsterdam. 408 pp.

LAWRENCE R.F. (1965). Some new or little known Solifugae from southern Africa. *Proc. Zool Soc. Lond.* 144, 47–59.

LEWIS T. (1960). A method for collecting Thysanoptera from Gramineae. *Entomologist*, 93, 27–28.

LUTZ H.J. & CHANDLER R.F. (1946). *Forest Soils*. Wiley, New York. 514 pp.

MACFADYEN A. (1961). Improved funnel-type extractors for soil arthropods. *J. Anim. Ecol.* 30, 171–184.

MACFADYEN A. (1962). Soil arthropod sampling. *Adv. Ecol. Res.* 1, 1–34.

MACFADYEN A. (1967). Methods of investigation of productivity of invertebrates in terrestrial ecosystems. In: *Secondary Productivity of Terrestrial Ecosystems*. K. Petrusewicz (ed.). Warsaw. 383–412.

McLEOD J.M. (1961). A technique for the extraction of cocoons from soil samples during population studies of the Swaine sawfly, *Neodiprion swainei* Midd. (Hymeoptera: Diprionidae). *Canad. ent.* 91, 888–890.

McNEILL S. & LAWTON J.H. (1970). Annual production and respiration in animal populations. *Nature, Lond.* 225, 472–474.

MILNE A., COGGINS R.E. & LAUGHLIN R. (1958). The determination of numbers of leatherjackets in sample turves. *J. Anim. Ecol.* 27, 125–145.

MURPHY P.W. (1962). *Progress in Soil Zoology*. Butterworths, London. 398 pp.

MURPHY P.W. (1962a). Extraction methods for soil animals. I. Dynamic methods with particular reference to funnel processes. In: *Progress in Soil Biology*. P.W. Murphy (ed.). Butterworths, London. 75–114.

NIERING W.A. (1963). Terrestrial ecology of Kapingamarangi atoll, Caroline Islands. *Ecol. Monogr.* 33, 131–160.

O'NEILL R.V. (1968). Population energetics of the millipede *Narceus americanus* (Beauvois). *Ecology*, 49, 803–809.

OOSTENBRINK M. (1954). Een doelmatige methode voor het toetsen van aaltjesbestrijdingsmiddelen in grond met *Hoplolaimus uniformis* als proefdier. *Meded. Landb. Hoogesch. Gent.* 19, 377–408.

PARIS O.H. & PITELKA F.A. (1962). Population characteristics of the terrestrial isopod *Arinadillidium vulgare* in California grassland. *Ecology*, 43, 229–248.

PĘTAL J. (1967). Productivity and the consumption of food in the *Myrmica laevinodis* Nyl. population. In: *Secondary Productivity of Terrestrial Ecosystems*. K. Petrusewicz (ed.). Warsaw. 841–849.

PETRUSEWICZ K. & MACFADYEN A. (1970). *Productivity of Terrestrial Animals: Principles and Methods.* Blackwell's, Oxford. 190 pp.

PHILLIPSON J. (1960a). A contribution to the feeding biology of *Mitopus morio* (F.) (Phalangiida). *J. Anim. Ecol.* **29,** 35–43.

PHILLIPSON J. (1960b). The food consumption of different instars of *Mitopus morio* (F.) (Phalangiida) under natural conditions. *J. Anim. Ecol.* **29,** 299–307.

PHILLIPSON J. (1962). Respirometry and the study of energy turnover in natural systems with particular reference to harvest-spiders. (Phalangiida). *Oïkos,* **13,** 311–322.

PHILLIPSON J. (1963). The use of respiratory data in estimating annual respiratory metabolism, with particular reference to *Leiobunum rotundum* (Latr.) (Phalangiida). *Oïkos,* **14,** 212–223.

PHILLIPSON J. (1967). Secondary productivity in invertebrates reproducing more than once in a lifetime. In: *Secondary Productivity of Terrestrial Ecosystems.* K. Petrusewicz (ed.). Warsaw. 459–475.

PHILLIPSON J. (ed.) (1970). *Methods of Study in Soil Ecology.* Unesco, Paris. 303 pp.

PHILLIPSON J. & WATSON J. (1965). Respiratory metabolism of the terrestrial isopod *Oniscus asellus* L. *Oïkos,* **16,** 78–87.

RAW F. (1967). Arthropoda (except Acari and Collembola. In: *Soil Biology.* A. Burges & F. Raw (eds.). Academic Press, London and New York. 323–362.

READ D.C. (1958). Note on a flotation apparatus for removing insects from soil. *Canad. J. Pl. Sci.* **38,** 511–514.

REICHLE D.E. (1968). Relation of body size to food intake, oxygen consumption, and trace element metabolism in forest floor arthropods. *Ecology,* **49,** 538–542.

REICHLE D.E. & CROSSLEY D.A. (1965). Radiocesium dispersion in a cryptozoan food web. *Health Physics,* **11,** 1375–1384.

REICHLE D.E. & CROSSLEY D.A. (1967). Investigation on heterotrophic productivity in forest insect communities. In: *Secondary Productivity of Terrestrial Ecosystems.* K. Petrusewicz (ed.). Warsaw. 563–587.

REYNOLDSON T.B. & YOUNG J.O. (1963). The food of four species of lake-dwelling triclads. *J. Anim. Ecol.* **32,** 175–191.

RICKER W.E. (ed.) (1968). *Methods for Assessment of Fish Production in Fresh Waters (IBP Hqndbook No.* 3). Blackwell, Oxford. 313 pp.

ROBERTS H. (1957). An ecological study of the arthropods of a mixed woodland with particular reference to the Lithobiidae. Ph.D. Thesis, Univ. of Southampton, U.K.

SAITO S. (1965). Structure and energetics of the population of *Ligidium japonicum* (Isopoda) in a warm temperate forest ecosystem. *Jap. J. Ecol.* **15,** 47–55.

SAITO S. (1967). Productivity of high and low density populations of *Japonaria laminata armigera* (Diplopoda) in a warm-temperate forest ecosystem. *Res. Popul. Ecol.* **9,** 153–166.

SAITO S. (1970). Methods for the study of production by macro-arthropods. In: *Methods of Study in Soil Ecology.* J. Phillipson (ed.). Unesco, Paris. 215–223.

SALT G. (1952). The arthropod population of the soil in some East African pastures. *Bull. ent. Res.* **43,** 203–220.

SALT G. (1955). The arthropod population of soil under elephant grass in Uganda. *Bull. ent. Res.* **46,** 539–545.

SCHØTZ-CHRISTENSEN B. (1965). Biology and population studies of *Carabidae* of the *Corynephoretum*. *Natura Jutlandica*, **11**, 1–168.

SIEGEL S. (1956). *Nonparametric Statistics for the Behavioural Sciences*. McGraw-Hill, New York. 312 pp.

SIMONSON R.W. (1957). What soils are. *Soils, Yearbook of Agriculture, Dept. Agric.* 17–31.

SMALLEY A.E. (1960). Energy flow of a salt marsh grasshopper. *Ecology*, **41**, 672–677.

SOLOMON M.E. (1957). Estimation of humidity with cobalt thiocyanate papers and permanent colour standards. *Bull. ent. Res.* **48**, 489–506.

SOUTHWOOD T.R.E. (1966). *Ecological Methods*. Methuen, London. 391 pp.

SPILLER D. (1948). Truncated log-normal and root normal frequency distributions of insect populations. *Nature, Lond.* **162**, 530.

STRICKLAND A.H. (1947). The fauna of two contrasted plots of land in Trinidad, British West Indies. *J. Anim. Ecol.* **16**, 1–10.

SUTTON S.L. (1968). The population dynamics of *Trichoniscus pusillus* and *Philoscia muscorum* (Crustacea, Oniscoidea) in limestone grassland. *J. Anim. Ecol.* **37**, 425–444.

SZABÓ I., BÁRTFAY T. & MARTON M. (1967). The role and importance of the larvae of St. Mark's fly in the formation of a rendzina soil. In: *Progress in Soil Biology*. O. Graff & J.E. Satchell (eds.). North Holland Publishing Co., Amsterdam. 475–489.

TODD V. (1949). The habits and ecology of the British harvestmen (Arachnida, Opiliones), with special reference to those of the Oxford district. *J. Anim. Ecol.* **18**, 209–229.

VARLEY G.C. (1970). The concept of energy flow applied to a woodland community. In: *Animal Populations in Relation to their Food Resources*. A. Watson (ed.). Blackwell, Oxford. 389–405.

VERHOEFF K.W. (1934). In: *Bronn, Klassen und ordnung des Terreiches* V/II/III/1–2 Lieferung.

WADLEY F.M. (1952). Notes on the form of distribution of insect and plant populations. *Ann. ent. Soc. Amer.* **43**, 581–586.

WALLWORK J.A. (1970). *Ecology of Soil Animals*. McGraw-Hill, London, 283 pp.

WHITE J.J. (1968). Bioenergetics of the woodlouse *Tracheoniscus rathkei* Brandt in relation to litter decomposition in a deciduous forest. *Ecology*, **49**, 694–704.

WIEGERT R.G. (1962). The selection of an optimum quadrat size for sampling the standing crop of grasses and forbs. *Ecology*, **43**, 125–129.

WIEGERT R.G. (1964). Population energetics of meadow spittlebugs (*Philaenus spumarius* L.) as affected by migration and habitat. *Ecol. Monogr.* **34**, 217–241.

WIGNARAJAH S. (1968). Energy dynamics of centipede populations (Lithobiomorpha— *L. crasspipes* and *L. forficatus*) in woodland ecosystems. Ph.D. Thesis, Univ. of Durham, U.K.

WILLIAMS R.W. (1960). A new and simple method for the isolation of fresh water invertebrates from soil samples. *Ecology*, **41**, 573–574.

YAGHI N. (1924). Application of the Roentgen ray tube to detection of boring insects. *J. econ. Ent.* **17**, 662–663.

Index

288